THE WORLD'S FIRST SHELL
COLLECTING GUIDE FROM 1821

"The Conchologists", a Victorian engraving depicting the lighter side of shell collecting in Georgian England.

THE VOYAGER'S COMPANION; OR SHELL COLLECTOR'S PILOT

The World's First Shell Collecting Guide
by John Mawe

compiled by
JEFFREY D. STILWELL

© Jeffrey D. Stilwell, August 2003.

The National Library of Australia
Cataloguing-in-Publication entry:

Mawe, John.
The voyager's companion; or shell collector's pilot:
the world's first shell collecting guide.

ISBN 1 920843 04 3.

1. Shells - Collection and preservation. I. Stilwell, Jeffrey D.
II. Western Australian Museum. III. Title.

594.1477

Published by the Western Australian Museum
Francis Street, Perth, Western Australia 6000
Edited by Jane Hammond Foster
Designed by Robyn Mundy
Printed by Lamb Printers Pty Ltd, Perth, Western Australia

FOREWORD

JOHN MAWE'S *Shell Collector's Pilot* — what an evocative title! It seems many years ago now that I purchased a copy of this early shell collecting guide in one of those grubby — but enticing — book grottoes that used to define London's Charing Cross Road. It was the fourth edition, published in 1825, and comprised a mere 82 pages and two hand-coloured plates between faded red boards. I read every word before completing the short rail journey home the same day.

A few years later some of the curious information in it re-emerged in a book I was writing about the history of shell collecting. The brief extracts I pillaged added some much-needed sparkle to the prosaic text of my own publication.

Now that Mawe's guide is reappearing in modern dress many more readers will be able to enjoy dipping into its pages, following its intrepid author along the coasts of North Africa and South America, sharing the pleasures of handling beautiful and rare shells gathered from shores lapped by the waters of the South Seas, learning how and where to collect shells and how best to preserve them.

Readers will discover that even unprepossessing shells are worth picking up, that the inhabitants of faraway places are often puzzled why anyone should want to collect such seemingly common objects, and that it is a good idea to pack shells securely in a box of sawdust or sand if they are to be transported a considerable distance. They might also be comforted by the thought that shell collecting can be financially rewarding.

Certainly they will want to look repeatedly at the two illustrative plates (while secretly regretting that they are not hand-coloured as they were in the copy I first read). When they read that many desirable species may be found on the coasts south of Rio de Janeiro, they will try to imagine what it would have been like to fish for Harps and Olives around Mauritius and Réunion, with a baited hook "and a line upwards of a thousand yards long".

I am pleased that Jeffrey Stilwell had the novel idea of reissuing this unassuming little work, delighted that he has added so much fascinating biographical and bibliographical information, and flattered that he has asked me to contribute these few words to it. John Mawe would be happy to know that this reprint of his *Shell Collector's Pilot* is likely to navigate its way onto the bookshelves of many collectors — of books and of shells — who may never have the opportunity to see, much less own, his own earlier editions.

S. PETER DANCE*
CAVENDISH HOUSE
CARLISLE, CUMBRIA, UK
18 OCTOBER 2002

---

* S. Peter Dance was former President and Honorary member of the Conchological Society of Great Britain & Ireland; and former President of the British Shell Collectors' Club. He is a popular lecturer on conchological subjects in the USA, appearing on television there as well as in the United Kingdom. He is a prolific writer on conchology, having published twenty books and over a hundred articles on molluscs and other subjects.

Contents

v | Foreword
S. Peter Dance

ix | Preface and Acknowledgements

1 | Introduction
The story behind the Shell Collector's Pilot

15 | A brief history of shell collecting

23 | The life of John Mawe (1766–1829)

33 | Facsimile of the first edition 1804
John Mawe

41 | Facsimile of the complete 1821
The Voyager's Companion; or Shell Collector's Pilot
John Mawe

79 | References and further reading

Published by J. Tennant late S. Mawe, 149, Strand.

WODARCH'S
INTRODUCTION
TO
THE STUDY OF CONCHOLOGY:
DESCRIBING

The Orders, Genera, and Species
OF
SHELLS:
WITH

OBSERVATIONS ON THE NATURE AND PROPERTIES
OF THE ANIMALS;

AND

DIRECTIONS FOR COLLECTING, PRESERVING, AND
CLEANING SHELLS.

BY

J. MAWE,

Author of Familiar Lessons on Mineralogy and Geology; Treatise on Diamonds and Precious
Stones; Travels in South America, and through the Gold and
Diamond Districts of Brazil, &c. &c.

FOURTH EDITION,

With considerable Additions and Alterations.

LONDON:
PUBLISHED BY LONGMAN, REES, ORME, BROWN, AND GREEN,
PATERNOSTER ROW;
AND BY S. MAWE, 149, STRAND.

1832.

One of John Mawe's last books, published by his wife, Sarah, three years after his death. (J D Stilwell Collection)

PREFACE AND ACKNOWLEDGEMENTS

SHELLS (phylum Mollusca) have a mystique that transcends time. If we delve into the abyss of the geological past, we find that shelled molluscs existed way back in the Early Cambrian Period, at the onset of the Palaeozoic Era, an amazing half billion years ago.

Thousands of years ago — a mere snip of time in this vast context — early humans began to appreciate the beauty and significance of shells. From the dawn of human history, people have combed beaches for nature's jewels of the sea and land — either answering an irresistible urge to discover perfect, beautiful specimens, or in the more noble cause of science.

My own passion for the history of shell collecting is rooted in my childhood in rural Indiana, United States. My great-great-grandfather, Ridgeway Wesley Stilwell (1842–1927), made a living from collecting freshwater shells along the Tippecanoe River and selling them to Indiana button factories. I could not help wondering who it was that first popularised the joys of shell collecting.

Many years later and quite unexpectedly, in a tiny bookshop in Cambridge, United Kingdom, I found the answer. Although I was on a research trip to study fossil Mollusca, my second passion is for rare book collecting. So I always make time when I travel to haunt the local bookshops in search of rare items.

I began, as I always do, by browsing the natural history book sections. Seeing nothing of interest, and hanging onto the remote possibility — as I always do — that there might a gem stashed away somewhere, I asked the proprietor about old shell books. He explained that the beautiful copy of Sowerby's *Conchological Manual* on the shelf was the only rare shell book in the shop — and it was one I already owned. But as I was leaving, he suddenly raced from behind his counter to stop me. He had completely forgotten about a little book he had priced only a few hours earlier. If I did not mind waiting, he would look for it in the pile of books behind the counter.

This had happened to me before, and I did not expect anything great or rare. But when he handed me the book, my curiosity was roused. Not only was it incredibly small but it was also very old and bound in half calf with marbled boards. My mind raced. In the countless libraries and book collections I had explored, I could not recall seeing a shell work of that minute size, particularly an old one.

The dealer told me that the book was frustrating him immensely as he could find absolutely no information on it and no book auction records of its sale for decades. It had been among an eclectic mix of oddities that he had bought from an eccentric elderly gentleman along the coast. He had to guess on a price, which, luckily, was very low.

I read the title on the leather spine — *The Shell Collector's Pilot* — and was none the wiser. I have a very good memory for mollusc works, but this was one I had never heard of. As I opened it I gasped at the hand-coloured frontispiece – tiny but exquisitely detailed. For such a small book, the title was bold and exciting: THE VOYAGER'S COMPANION; OR SHELL COLLECTOR'S PILOT.

The book that I had discovered had been almost forgotten — the third edition of a little gem on shell collecting by John Mawe.

At present I am one of only a handful of people now holding copies of both the third and fourth editions of this book in private hands — and these copies are considered to be the best in existence. Some 14 examples of the 1825 (fourth) edition have been traced, and of these only a few are owned privately.

This is why I decided to reproduce the book here — so that many more people could enjoy one of the greatest rarities of early nineteenth-century natural history. Readers will discover that while the techniques and joys of shell collecting have changed little, our knowledge of the Mollusca has increased in great leaps and bounds to the present day.

Many people have assisted me throughout the course of this project and have made the re-publication of *The Voyager's Companion; or Shell Collector's Pilot* a very enjoyable task.

The project would have never come to fruition had it not been for the enthusiasm and expertise of Ann Ousey, Western Australian Museum Press, Perth. Erudite shell collector and prolific author S. Peter Dance, of Cavendish House, Carlisle, UK, sent details of Mawe's publications, and photocopies of original holographs in his collection, written by Sarah and John Mawe. Mr Dance also kindly consented to write the Foreword to this edition of Mawe's book. Professor Hugh Torrens, University of Keele, UK, brought the exceedingly rare first edition (1804) of Mawe's book to my attention, and contributed valuable comments on the text and additional unpublished information from his on-going research on Mawe.

I wish to acknowledge and thank the Mitchell Library, State Library of New South Wales, for agreeing to the duplication of the 1804 edition for the purposes of this book. I owe a great deal to the friendly and helpful staff at the Natural History Museum in London — Carol Gokce, Lorraine Portch, Dr David Reid, Vicki West, and others — who chased up and copied bibliographic information on Mawe and his holographed letters in the Museum collections. Carolyn Tredrea and Rita Bisley, of the James Cook University Library, Townsville, Queensland, helped to track down existing third and fourth edition copies of Mawe's book in world collections.

Brian Pump, of Townsville, provided invaluable photographical assistance throughout the production of this book. Dave Cossey and Eddie Thornton, of NadicPrint Services Pty Ltd, Townsville, created the digital images of Mawe's book and other shell images used in this publication. Adella Edwards of James Cook University scanned images as well.

Others who have provided information and/or much needed support include Dr Fred Wells and Dr Ken McNamara, Western Australian Museum; Dr Peter Middelfart and Dr Winston Ponder, Australian Museum in Sydney; Jenny Broomhead, Maria Burns, Maggie Patton and Josef Stejskal, State Library of New South Wales; Richard Petit of North Carolina; Dr Patty Jansen of Sydney; and Jeffery Rodda of Townsville.

A prolific author in his own right, my colleague and friend, palaeontologist Dr John Long of the Western Australian Museum, encouraged me to publish this historical and special book and provided me with details of his long experience in publishing popular works for the public. His passion for natural history has inspired me to present this rarity again to the shell-loving community.

INTRODUCTION
The story behind the Shell Collector's Pilot

Two centuries ago, there was one lover of shells in England, who stands head and shoulders above the crowd. A vivacious "go get 'em" sort, he turned shell collecting into a science — and a very lucrative business.

This was John Mawe (1766–1829), who travelled the globe in search of the most prized and beautiful shells and minerals, and paid scouts handsomely to send his specimens to England.

Mawe became a dealer in natural history objects and wrote several important books on conchology (the study of molluscs or shelled invertebrates), as well as mineralogy and allied subjects. Over many years, he and his wife, Sarah, handled some of the most valuable and significant shell collections, including the famous Captain Bligh collection, which passed into their hands upon his death.

Mawe's ground-breaking little book, *The Voyager's Companion; or Shell Collector's Pilot: With Instructions and Directions where to Find the Finest Shells…*, jointly published in 1821 by the author and Longman in London, was priced at a mere five shillings — reasonably affordable to those of modest means.

In this first detailed and all-encompassing account of shell collecting, Mawe carefully outlines the proper way to collect choice specimens, and reviews the best places to find interesting shells in all parts of the world at that time. He also explains how to collect birds, reptiles, quadrupeds, plants and minerals, but the focus is on shells.

Interestingly, the techniques of collecting specimens have changed little over time. The early collectors, including Mawe, agreed that the shells should be washed well in fresh water, but the uninformed would "often grind down the fine epidermis of rare shells, on the voyage home, to make them look pretty". Scientists studying Mollusca in Georgian England would pay more for a shell covered by its rough coating (a skin or horny covering on the exterior of many shells, technically called the "periostracum") than when it had been taken off by unskilled hands.

Mawe advised that, for the long journey back to England, shells needed to be packed in sawdust — or sand as a second choice. He stated firmly that, after sealing the barrel or chest with the shells safely and correctly packed inside, the container should be "properly marked, and entered into the *Ship's* manifest, as SHELLS — to prevent seizure".

In the following letter, written in Mawe's hand to Robert Saunders, of St Pauls Street, Exeter, and dated about 1821, he outlines the proper procedures for preserving molluscs:

Mr Mawe's Method of Cleaning Shells

Dissolve half a pound of Pearl ash [sic] and half a pound of soft soap in two
quarts of boiling water stirring it until all the particles are dissolved — pour it warm
over the shells, let them remain in this two or three days frequently warming it and
pouring it over them, rinse them out of this with a brush and clense [sic] them well
in warm water, when dry brush them with a nail brush, and with a bit of cotton in
oil and rub them with it, then clean them with a soft leather and linnin [sic] rag and
if they are live fine shells they have a sufficient pollish [sic] ———————
Rugged shells such as muscles [sic] often require a different process though it is
prudent to try that method first, but when the shells are covered with adhesions or
the epidermis will not seperate [sic] it is then necessary to use acid which should
always be applied by a carefeul [sic] hand after the shells have undergone the
foregoing process and are perfectly dry with a small brush such as the painters use
for painting window frames, apply strong muriatic acid to the parts requiring it

dipping the brush in sand and using constant friction until the epidermis and adhesions are removed, after this clense [sic] them in warm water and dip them in alkali to neutralize the effects of any acid that may remain, then cleanse them in warm water again, and when dry with a soft brush put a weak solution of gum arabac [sic] over them. It is sometimes requisite to seperate [sic] the epidermis and adhesions with a strong pen knife

    The pollished [sic] shells such as the *Haliotis*, &c., are ground and pollished [sic] with a Laythe [sic].

    NB Shells should never be dipped or lie in acid as it inevitably injures the enamel of the shell.

ORIGINAL IN COLLECTION OF S. PETER DANCE, CUMBRIA, UK

Even nondescript and rather ugly-looking shells that were prolific in their native countries were often considered valuable back in England. Mawe tells in his book how he would become frustrated by repeated attempts to obtain shells from some areas because the locals could not fathom why on earth he would want them. "I have found it difficult to persuade any one to send me another supply," he writes, "although so easily obtained. No, they think them common, and because they *really are so there,* they do not bring or send them."

    The ability to differentiate between common and rare shells — "select the good from the bad" — would only come with long-term practice, Mawe explains, and by handling many varieties. He also implies that this skill could have monetary rewards in the end and should not be taken lightly.

    The *Shell Collector's Pilot* is a book of Lilliputian proportions at approximately 8.5 x 14.0 centimetres, with xv + 56 numbered pages. It is literally one of the smallest books ever written on shells and is printed in incredibly tiny, eye-straining text. Its size probably reflects Mawe's intention for it to be a field guide, something that was rare for the time. Only *Pinnock's Catechism of Conchology*, published in London in various editions during the 1820s, is of a similar size at 8.8 x 13.7 cm, but this is not a field guide.

    The hand-coloured, copper-engraved frontispiece (6.2 x 9.7 cm), apparently drawn by Mawe himself (his name appears directly under it) and delicately painted in watercolours by

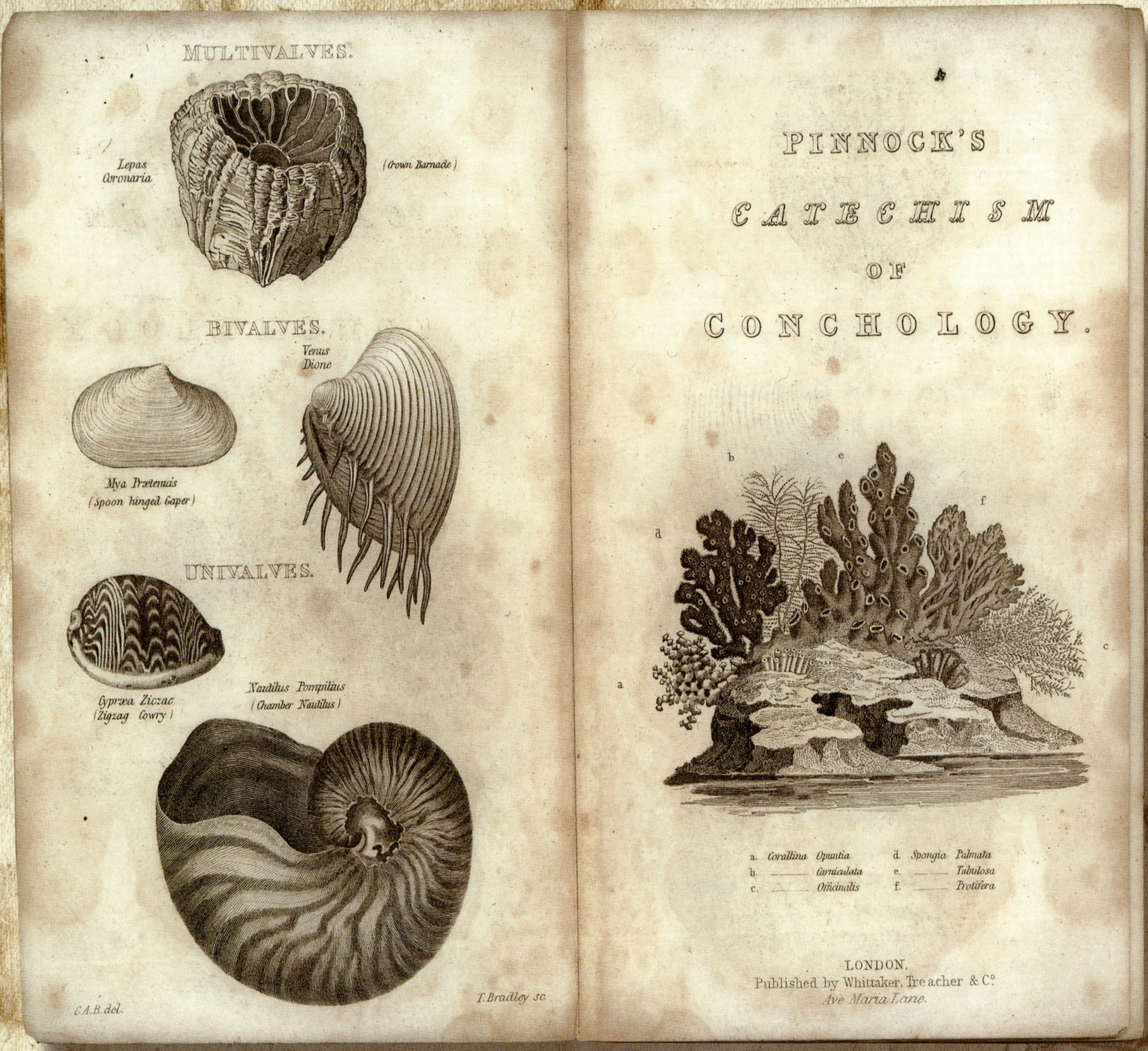

PINNOCK'S
CATECHISM
OF
CONCHOLOGY.

a. *Corallina Opuntia* d. *Spongia Palmata*
b. ———— *Carniculata* e. ———— *Tubulosa*
c. ———— *Officinalis* f. ———— *Protifera*

LONDON.
Published by Whittaker, Treacher & C°.
Ave Maria Lane.

Pinnock's Catechism of Conchology, *published in London in 1824 (2nd edition 1829), is one of the smallest books ever written about shells (8.8 x 13.7 cm). (J.D.Stilwell Collection)* a skilled hand (probably Sarah Mawe), is arguably one of the most beautiful depictions of natural history objects in Georgian England. It illustrates palm trees and tropical birds, including Australia's sulphur-crested cockatoo, with brightly coloured shells strewn on a beach — a shell collector's paradise. Underneath is the appropriate caption, "Search and ye shall find", and directly below this, "Oh! Qué rara couza!" This is Portuguese for "Oh! What rare things!" and alludes to Mawe's delight in the discoveries he had made in Brazil. An original of the copper engraving survives in the Geological Society Library, UK.

In his own words, Mawe uses this frontispiece to "tempt the traveler, and rouse his mind to contemplate on the beauties of the deep, and the wonderful words of an Omnipotent Being". By including such a detailed, breathtaking — and no doubt expensively produced — illustration, together with the coloured plates, Mawe hoped to inspire more people to search for these stunning objects, which would in turn benefit his dealership.

To reduce the cost of producing lavish books with rich, brightly coloured illustrations, as well as for convenience, it was quite common at that time for female family members and friends to assist authors by hand-colouring copies of their books. The variation in hand-colouring (from soft to very bright) of the plates in copies of the *Shell Collector's Pilot* and Mawe's other books suggests that girls and/or women other than Sarah Mawe did some or all of the colouring.

The 1821 edition is considered to be a third edition; the first two were quite different, having not only a different title, but different content and no illustrations. A fourth edition was published in 1825 at the same price as the third — a modest five shillings — and is a slightly expanded version of the 1821 edition, with some 20 pages added giving further information on worldwide shell localities. It is also slightly larger in format by about 15 per cent and includes a one-page advertisement. This expanded edition perhaps reflects the burgeoning interest in shells during that four-year period.

Mawe refers in the *Shell Collector's Pilot* to the book having gone through two editions "several years ago". Those first two editions were entitled *A Short Treatise, Addressed to Gentlemen Visiting the South Seas, and All Foreign Countries; More Particularly to Commanders, &c. of Ships, and Gentlemen Residing on Shore, With a View to Encourage the Collecting of Natural History*.

The only copy (possibly unique) of the first (1804) edition that has been traced worldwide is in the State Library of New South Wales, Sydney, and was brought to my attention by Hugh Torrens of Keele University, UK. It has not yet been included in any global database. As far as we know, there are no surviving copies of the second edition at all. There were probably no more than 50–75 copies of the first two editions printed, and perhaps 100 of the third and fourth.

As S. Peter Dance has pointed out, the first two were mostly likely unbound and therefore quite easily mutilated, destroyed or lost during voyages. And the third and fourth would have suffered a similar fate. In all there are only nine surviving copies of the book that we know about. This is probably why it hardly gets a mention in histories of shell collecting, despite being described by Dance as "The fullest and most entertaining shell collector's guide".

The book has been produced with three kinds of binding: — finely pebbled cloth, plain paper-covered boards (both with printed paper label containing title), and more expensive half calf with marbled boards and morocco gilt spine label with title.

The first edition (reproduced here), which was really little more than a pamphlet of just 11 pages, appears to be Mawe's third published work after *The Mineralogy of Derbyshire* (1802) and *Catalogue of Minerals* (1804). Presented here for the first time are bibliographic details of these early editions, which were printed by Young at Brydges Street, Covent Garden. Mawe's shop was nearby at 5 Tavistock Street, Covent Garden.

Bibliographies include a different title, *Directions to Captains of Ships, Officers, and Travellers, Particularly Those Who Visit the South Sea Islands…*, and without any publication details. This is the title that Mawe mistakenly gives to the earlier editions in his *Shell Collector's Pilot*. He was probably relying on his memory and had published too many books by this time to recall all the titles accurately.

Mawe begins the first edition by discussing the various uses of molluscs, such as obtaining dye from *Purpura,* and the medicinal importance of certain mollusc animals in treating ailments, such as pulmonary complaints. In a style commonly found in his later books, Mawe ends the first page by suggesting that he "will amply repay those who take the Pains to collect them".

He also includes an advertisement of his trade in minerals: "His Cabinets exhibit above 20,000 Varieties, which are always open for Public Inspection." At this time he specialised in

A BRAZILIAN MINER WASHING THE ALLUVIAL SOIL (RAKED FROM THE RIVULET) FOR GOLD & DIAMONDS.

VIEW NEAR MATLOCK, DERBYSHIRE.

FAMILIAR LESSONS ON MINERALOGY AND GEOLOGY:

WITH

COLORED PLATES.

To which is added

A PRACTICAL DESCRIPTION OF THE USE OF

THE

Lapidary's Apparatus,

Explaining the Methods of slitting and polishing Pebbles, &c.

BY JOHN MAWE,

Honorary Member of the Mineralogical Society of Jena, &c. &c.
Author of Travels through the Gold and Diamond District of Brazil; Treatise on Diamonds and Precious Stones, New Descriptive Catalogue, &c.

Whose hand unseen the works of nature dooms,
By laws unknown! WHO GIVES AND WHO RESUMES.

SIXTH EDITION.

London:
PUBLISHED BY THE AUTHOR, 149, STRAND;
AND LONGMAN, HURST, REES, ORME, BROWN, & GREEN, PATERNOSTER ROW.
1824.

First published in 1819, Mawe's Lessons on Mineralogy and Geology *was immensely popular and went through many editions throughout the 1820s and early 1830s. (J D Stilwell Collection)* minerals more than shells, but later in his career this changed, and he eventually bought and sold immense collections of shells.

The following comment in the *Pilot*, on packing shells for transport back to England, conjures up images of long past travel by ship:

> One Remark, not the least worthy of Notice, is, that valuable Shells are frequently spoiled after they are collected, by being put into empty Beef Tierces [a cask or vessel containing an old measure of capacity equivalent to one third of a pipe, or 42 wine gallons], &c. without packing, where they rub one against another as the Ship rolls, by which means their Characters are entirely defaced and spoiled. It cannot be too strongly recommended to pack them in a Box, and nail them down as soon as they are collected.

Mawe's third edition attempts to persuade captains of ships and travellers to follow his instructions carefully on how to avoid problems with customs officials when collecting shells. All would be made simple, he explains, if the correct procedures were followed to the letter. If collectors would send their shells, packed in cases, to him c/o King's Warehouse, London, they would not be seized by customs officers if entered on the ship's manifest. Mawe would then pay both for the shells and the "trivial" customs charges. For shells imported into Britain, these charges amounted to 20 per cent of their value, until the tax was repealed in July 1825.

Mawe repeats his request for shells to be collected and sent to him in various places throughout his books. He was clearly an expert dealer. For many years, he spent an average of £200 or more a year — an enormous amount at that time — for shells from the Australian region, although specimens from other regions were even more highly valued. His own immense collection included rare volutes and land and freshwater varieties, from the collection belonging to the Dutch Governor of Ceylon, taken in 1795. An astounding £10,000 was paid for shells from Ceylon between c.1810 and 1820. More than two-thirds of them passed through Mawe's hands.

But he was less of an expert when it came to locality information. Accurately recording the place where specimens were found was quite a problem at that time. Mawe appears to have been either too credulous or, more likely, careless about this, as many of his specimens are apparently from what we now know are impossible locations. From this, S. Peter Dance concluded that large areas of the globe were, conchologically speaking, unknown in Mawe's day.

To be fair, this was an age of collecting for the "cabinet" rather than for science. Even as late as 1820 the value of the shell was probably still measured in terms of rarity and aesthetic beauty, so that the accuracy of the documentation of various species was deemed rather unimportant. A personal letter, probably written by Mrs Mawe, includes the following:

> Scarcity in Shells in the opinion of genuine collectors often outweigh beauty. The large *Cypraea Tygris* which is very handsome is exceeded in value by the trivial looking *Cypraea punctata* and which is seldome [sic] met with even in valuable collections. *Cypraea Aurosa* or Orange Cowrie is a rare and magnificent shell and is worn as a mark of Royalty by the South Sea Islanders …
>
> S. Peter Dance collection, Cumbria, UK

So the shells sold by dealers in Great Britain and Europe were often incorrectly localised, in terms of region, and sometimes even country or continent. As a result, many student and researcher met major stumbling blocks in documenting the geographic range of a species, and no doubt there were many wild, global goose chases for particular shell rarities.

For example, Mawe states in the 1825 edition that "hammer oysters [*Malleus* species] and beautiful cones" had been brought from New Zealand, but the *Conidae* had been extinct in this region for some five million years. And he had obtained, apparently from the Philippine Islands, the "Junonia [*Scaphella junonia* Shaw]", but this species, which seems to have been exceedingly rare in pre-Victorian times, is found today only along the coast of Florida.

Dance points out that Mawe summarised in detail in his books the most up-to-date geographical information of the day. "We are little acquainted with the shells from the south coasts of the Mediterranean, Malta, Sicily or the [Greek] Archipeligo", he writes, but from Mogador he had received "fine limpets … muscles, and various land helices, which I persuaded the Moors to bring from the interior".

A TREATISE ON DIAMONDS
AND
Precious Stones,

INCLUDING THEIR

HISTORY—NATURAL AND COMMERCIAL.

TO WHICH IS ADDED,

SOME ACCOUNT OF THE BEST METHODS OF
CUTTING AND POLISHING THEM.

BY JOHN MAWE,

AUTHOR OF TRAVELS THROUGH THE DIAMOND DISTRICT OF
BRAZIL, &c.

London:
PRINTED FOR LONGMAN, HURST, REES, ORME, AND BROWN,
PATERNOSTER-ROW.

1813.

Mawe's A Treatise on Diamonds...(1813), the first coloured plate book on these precious stones. (J D Stilwell Collection)

CONCHOLOGY.

with the *Cardia* and *Tellinæ*; to the former they are more closely allied; it was originally established by Scopoli under the name of *Sphærium*, this was changed by Lamarck to *Cyclas*, which is now adopted by Latreille and other writers, notwithstanding the previous application of the term to a genus of plants.

SPECIES.

1. *C. Similis.* Shell suborbicular convex, base a little flattened; with nearly equi-distant, raised, concentric lines, giving a slightly sulcated appearance to the surface, and generally a more conspicuous elevated darker wave, marking the former year's growth of the shell. Epidermis brown or ferruginous; beak nearer central and obtuse; hinge with minute very oblique teeth, lateral ones very distinct, elongated, and considerably resembling those of the next species.

Length seven-twentieths of an inch: breadth two-fifths: a specimen measured in length nearly three-fifths of an inch.

Plate 1. fig. 9.

Very much resembles *Tellina Cornea* of authors; is found in plenty in the river Delaware; animal viviparous; from one specimen three pale yellow active young ones were taken, the largest of which measured in breadth three twentieths of an inch in the month of May.

2. *C. Dubia.* Shell oblique, subovate convex, concentrically wrinkled, very pale horn colour or whitish, with sometimes a darker, but not raised band, marking the preceding year's growth of the shell; beaks placed much nearer one end; within whitish, primary teeth very distinct, in one valve two divaricating ones, in the other but one, exterior lateral laminal tooth very small.

Length, five-twentieths of an inch; breadth, three-tenths.

Plate 1. fig. 10.

Inhabits the river Delaware in company with *C. Similis*, and very much resembles Tellina Amnica of Authors.

GENUS CYRENA, *Lam.*

Shell triagonally rounded, turgid, inequilateral, equivalve, robust; umbo eroded or decorticated; hinge three toothed in each valve; lateral teeth two, of which one is placed nearer to the primary ones; ligament exterior placed on the longest side.

Obs.. This genus has been but lately constructed by Lamarck, to receive such shells of the former genus *Cyclas* as have three primary teeth on each valve.

SPECIES.

C. Caroliniensis. Shell cordate, turgid, brown on the disk, with a yellowish or greenish margin and sub-margin, surface with numerous membranaceous wrinkles; umbo much eroded; beaks distant; two of the primary teeth caniliculate at tip.

Length, one inch and one-fourth; breadth, one inch and seven-twentieths.

Cyclas Caroliniensis, Bosc.

Inhabits the rivers of South Carolina, and Georgia, but is not found so far north as Pennsylvania. We found it in plenty near Charleston, South Carolina, and in St. Johns' River, East Florida.

The shells here described are in the collection of the Academy of Natural Sciences of Philadelphia.

It was originally the intention of the writer of this article to insert here, not only descriptions of the fresh water and land shells, but those of the coast also; finding, however, that the descriptions of the latter were by far too voluminous to be comprised within the space allotted to this article, and that they had more generally found a place in the systems, the design is, with respect to this work, necessarily relinquished. To all the species here described, with the exception of three or four, we have been constrained to adapt specific names; but should it appear that we have been anticipated by the labours of some recent conchologist, whose writings we have no opportunity to consult, we shall readily bow to the right of priority, which ought unquestionably to be on all occasions imperative and exclusive.

The primary divisions of the Linnæan system, in the latest edition of the "Systema Naturæ," as before observed, consist of three orders, Multivalve, Bivalve, and Univalve, each of which is subdivided into genera. The Multivalves contain the chiton, leapas and pholas: the Bivalves, mya, solen, tellina, cardium, mactra, donax, venus, spondylus, chama, arca, ostrea, anomia, mytilus, and pinna; and the Univalves, argonauta, nautilus, conus, cypræa, bulla, voluta, buccinum, strombus, murex, trochus, helix, nerita, haliotis, patella, dentalium, serpula, teredo, and sabella. Which see. See also SHELLS.

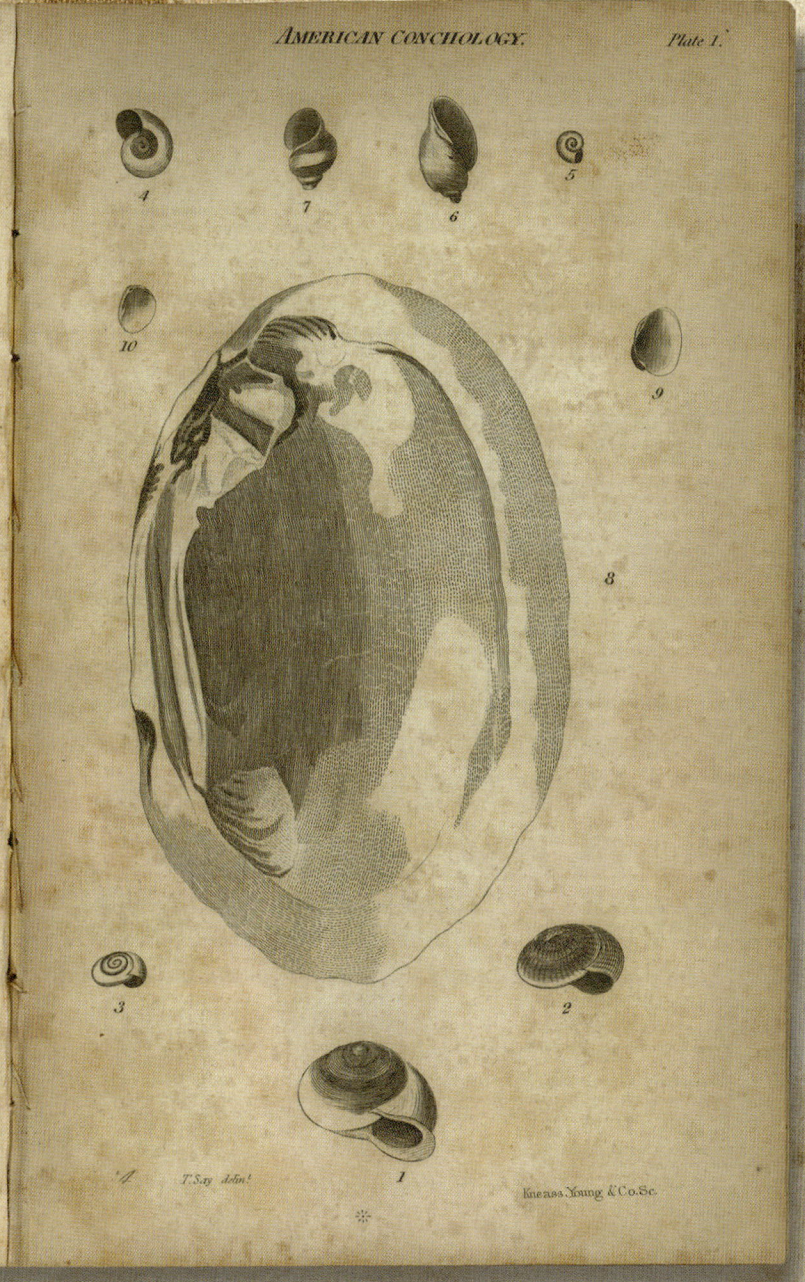

An extremely rare copy of Thomas Say's (1787–1834) monograph on shells (1819, 3rd edition), the first work on American molluscs. (J D Stilwell Collection)

He had little positive to say about shells from the West Indies, stating with a slight air of arrogance: "The collections of shells, in fine boxes, which are sold in the West-Indies, are too contemptible to notice; they can only please mere novices."

With regard to Africa, Mawe writes: "… the negroes who trade to Sierra Leon, have been induced to bring land shells from the interior; but, as they often apply fire to the shell, in order to drive the animal out, they are thereby much injured, and often entirely spoiled". However, he recommends hiring "a fisherman, or clever negro, (who ought to be well remunerated on his return), to collect, as well as to pack these objects, and thus free the employer from any trouble".

Scarce information was available on the Congo or the interior, but many fine specimens had been found in the Portuguese settlement at Benguela, and the coasts on each side of the Cape of Good Hope abounded with interesting shells.

The North American coasts are home to "some interesting varieties", although it appeared that cold climates in general "were unpropitious to the production of richly-coloured shells". The coasts of Georgia, and North and South Carolina, produced many beautiful species, particularly of *Oliva*, but "of the testaceous productions of the island of New Orleans, the Mississippi, and the whole of that range, we have obtained little information".

The data that had been collected to date in North America came from the early research of naturalist Thomas Say. In 1816, Say had founded conchology in America by publishing the first treatise on American shells, *Descriptions of Land and Fresh-water Shells of the United States*, an extremely rare 15-page paper published in Philadelphia with four plates of terrestrial and freshwater clams and snails.

Very rare and beautiful shells were taken from remote areas as well (or areas that are remote from Europe). Mawe collected many in Brazil, especially from the coasts south of Rio de Janeiro. The finest volutes to reach England were collected on "Bligh's Island" in the Fiji group. And a friend of Mawe's obtained cowries from Guam, such as "the most beautiful *Cypraea aurora* [*Cypraea aurantium* Gmelin] that has yet appeared".

From the Philippine Islands he received the Imperial Volute (*Aulica imperialis* Lightfoot), "and many other varieties of great rarity and beauty". He also secured shells taken from relatively deep water around the islands of Mauritius and Réunion, where officers and men stationed there employed much of their leisure time "in fishing for Harps and Olives, which they effect by means of a baited hook, and a line upwards of a thousand yards long".

Beautiful gastropods were found in Australia too, and the finest *Phasianella australis* Gmelin ever brought to England were gathered by two boys in Western Port, Victoria:

> A whaler [whale ship] off the coast sent a boat on shore to search for fresh provisions, as birds, animals, &c. whilst the crew were shooting, the boat grounded amongst stones and weeds, and during the time before she floated, the boys left in charge of her, employed themselves in gathering the shells entangled in the weeds and about the stones [from whence they took many limpets and large chitons – 1825 edition]. On their return home they brought them to me; and though I gave them the price they asked, I thought it not only right, but political, to present them with a guinea each, in order to stimulate them to look out for shells on another voyage.

Mawe lamented that "it is, generally, only the boys or the cook, who notice these *rarities*, and who make a few pounds by them every voyage".

He appropriately concludes his discourse on shell collecting by simply telling the reader to "keep a *good look out*".

13

CONCHOLOGY. Plate IV

1. The Dipper, or Pewits Egg, and all in the circle are back Views of the same. 2. The Argus Cowry, and all in the middle and corners are different species of Cowries. 3. The Weavers Shuttle.

Albertus Seba del. J. Chapman sculp.

London Published as the Act directs Aug.st 24.th 1802. by J. Wilkes.

A BRIEF HISTORY OF SHELL COLLECTING

THE WORLD'S FIRST shell collectors were early humans, who discovered that molluscs were a rich, reliable and tasty source of protein — and much easier to catch than swift-moving animals.

People who lived in coastal areas had near-endless supplies of scrumptious mussels, oysters, scallops, cockles and snails. Their daily feasts led to an important record for achaeologists, as discarded shells eventually became huge mounds, or "middens", that rose higher and higher with each generation.

As incredible as it may seem, some shell middens became so high and deep — such as those found in the Orkney Islands — that the inhabitants had to form tunnels through the middens to get from one place to another.

The middens made by these early peoples are to be found in coastal areas of nearly every corner of the globe, — following the shorelines of ancient seas, to inland areas of long-vanished lakes and river beds, where freshwater molluscs once flourished.

Left: A beautiful plate of cowries (Cypraeidae) from a later edition of Albert Seba's (1665–1736)'s Thesaurus, *dated 1802. (J D Stilwell Collection)*

I have recorded early shell middens from many areas in New Zealand and the remote Chatham Islands made by the Maori and Moriori peoples, and also in southern South America, where Charles Darwin wrote of many tons of shells in mounds in Tierra del Fuego.

As early as 40,000 years ago, in Sarawak (Borneo), the charred remains of snails suggest that the early inhabitants roasted these molluscs before they were eaten. In Australia, early Aborigines amassed hills of broken shells covering nearly 0.2 hectares and three metres deep, and on top of the mounds were remains of fireplaces where the shells had been cooked. Heaps of uncooked shells were found beside the fireplaces, apparently abandoned at the last second for some mysterious reason.

Although these early shell collectors clearly had food on their minds rather than aesthetic beauty, there is evidence from late Palaeolithic times that some shells may have been used to adorn the bodies of both men and women. This most likely signified an innate, primitive appreciation of the beauty of nature and its products. But we will never know exactly why some early burials also include shells. — What mystical purpose did they possess? Perhaps they were a gift of food for the next life.

Archaeologists have found marine shells in ancient graves all over the globe, some more than 15,000 years old. These include a Pacific shell discovered in the ruins of an Arizona pueblo, a North Sea shell in a Swiss lake village, Mediterranean and Indian Ocean shells in the ashes of Pompeii, and an Atlantic shell in an Etruscan grave. These shells were probably taken as novelties on various early voyages that focused on trade.

Shell collecting for its own sake reputedly began with a madman — the Roman Emperor Caligula (original name Gaius Caesar), who ruled from 37 AD until his death four years later. Having arrived with his legions at the English Channel during the spring of 40 AD, he looked at the formidable barrier and wrote for posterity that he would rather conquer Neptune than the Britons. So he issued a bizarre order: he commanded his troops to form a battle array and to start collecting shells along the French shoreline. Caligula was then able to return to Rome with what he called "the spoils of conquered ocean".

Some of the first serious voyages for shells were undertaken by the ancient Phoenicians, who travelled as far as the British Isles and circumnavigated Africa to seek out new beds of dye-producing murex gastropod shells, such as *Bolinus brandaris* Linnaeus and

Purpura patula Linnaeus. These snails were used in the manufacture of the Tyrian purple dye that was important from Roman times to the Middle Ages for ecclesiastical and imperial robes.

Other voyages were sponsored to find pearl shells and money cowries (cypraeid gastropods). The latter had been popular since ancient times, as the aperture resembles the female genital opening and perhaps assumed perceived magical powers.

Little happened on the collecting scene until Dutch merchant ships sailed back to Amsterdam with the colonial spoils of the East Indies during the seventeenth century. It was a tradition at this time for cultivated and erudite people — mostly men — to assemble "cabinets", which were generally large rooms packed with every kind of curious and natural object imaginable: — animals, birds, fish, shells, minerals, weapons, primitive tools, coins, bones …. It was also fashionable, during the early eighteenth century, to display a glass pyramid full of shells that was sometimes topped by a beautifully preserved pearly *Nautilus* shell.

The British Museum collection in London was formed around the nucleus of the great collector Sir Hans Sloane's cabinet, begun in the 1600s. In 1625 Englishman John Tradescant commissioned Edward Nicholas, on behalf of the Duke of Buckingham, to "Deall with All Marchants from All Places But Espetially the Virgine & Bermewde & Newfound Land Men … to furnishe His Grace With All manner of Beasts & fowells and Birdes … shells … Bones Egge-shells …". This later became the cornerstone of the Ashmolean Museum collection in Oxford.

Although the collector's cabinet has long gone out of fashion, surviving no later than the early 1800s, serious collectors still exist today, who fill their houses with precious treasures in cases.

At the end of the seventeenth century, two major figures in the early history of conchology came onto the stage: the Englishman Martin Lister and the Italian Jesuit, Philippo Buonanni (or Bonanni). These two men, together with Georgus Everhardus Rumphius (discussed below) are the pioneers of the wonderful world of shells. Lister wrote the famous book *Historiae animalium Angliae tres tractatus* …, published in 1678, which is a detailed work on spiders and shells, and also includes fossil forms. His extensive, most well

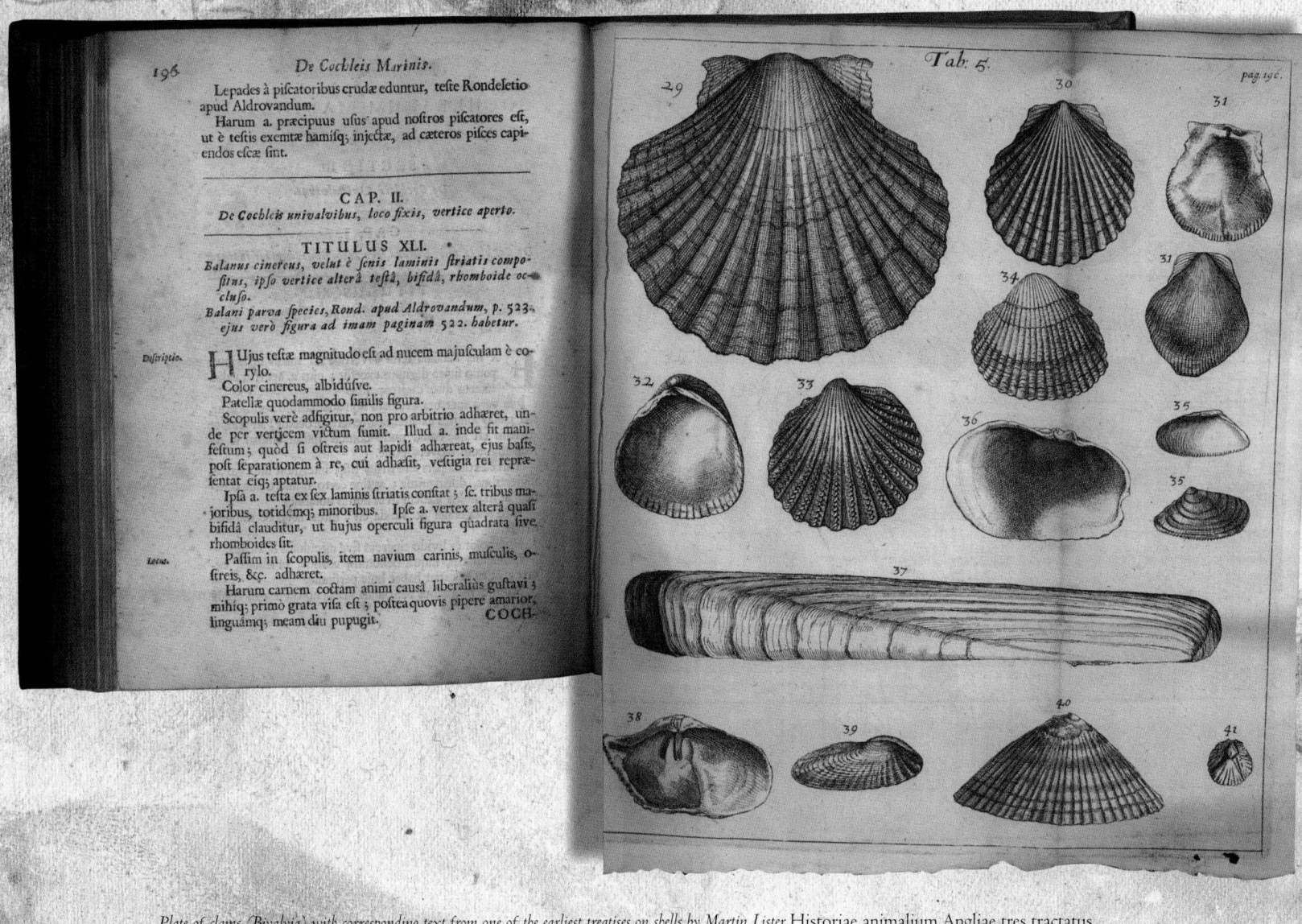

Plate of clams (Bivalvia) with corresponding text from one of the earliest treatises on shells by Martin Lister Historiae animalium Angliae tres tractatus... *published in London in 1678. (J D Stilwell Collection)*

Double-page frontispiece of:
Thesaurus Imaginum Piscium Testaceorum *(1739) by Georgus Eberhardus Rumphius (1627–1702).*

The only known portrait of Rumphius drawn from life, showing the blind author in his study surrounded by his natural history treasures and "cabinets". (J D Stilwell Collection) known work, *Historia conchyliorum*, followed in 1685–92, going through several editions. Buonanni, however, was the first to publish a book that was entirely devoted to shells, — *Ricreatione dell' Occhio e della Mente,* — in Rome in 1681. The engravings, although well done, are not considered to be top rate by most conchologists, but the book is highly sought after nevertheless. As the title suggests, — *For the Delight of the Eyes and the Mind,* — it was a very expensive production for the time.

One of the greatest collectors of all time was George Eberhard Rumpf (or Georgus Everhardus Rumphius) (1628–1702), who had entered the services of the Dutch East India Company in 1652. He went first to Java and later to the Moluccas. After he became blind in 1670, the author stayed on the island of Ambon, mainly dedicating himself to the writing of his publications. In 1701 he sent the manuscript of his *Thesaurus Imaginum Piscium* to his friend and mayor of Delft, Hendrik D'Acquet, who arranged the publication through the Amsterdam publisher François Halme. His work is full of accurate and detailed information, focussing on molluscs. It is presumed that its excellently engraved plates are the work of Simon Schynroet or Johan Philip Sipma, done after drawings by Maria Sibylla Merian. Rumphius put together large collections that became the basis of one of the classic books on shells, *Amboina Cabinet of Rarities*, written in 1705. The great Swedish naturalist Karl Linné (or Linnaeus) used some of Rumphius' shell names in his *Systema Naturae*, published in 1758 (10th edition), and these names are still used today.

At about the same time, another incredible collection was being formed. The wife of Captain William Bligh based her collection on gifts brought home by her husband, whose ship, the famous *Bounty*, took him on many voyages to the Pacific. There he collected some of the rarest shells that had yet been seen by Europeans. The many treasures — which passed through the hands of John and Sarah Mawe and eventually made their way to the British Museum — included a precious wentletrap, the epitoniid gastropod *Epitonium scalare* Linnaeus (literally meaning "spiral staircase").

Today this gastropod spans the sea from Japan to northern Australia, but it was so rare in the nineteenth century that enterprising Oriental craftsmen supposedly made counterfeit specimens from rice-flour paste, which, of course, fell to pieces after washing! It is astounding to think that a nice specimen of the wentletrap was purchased in 1750 by Emperor Francis I of Austria for today's equivalent of many tens of thousands of dollars.

Thankfully the price has dropped substantially since then to only a few dollars, so most people can afford to add this beautiful gastropod to their own collections.

With these voyages, the collections in Europe flourished with the most rare and beautiful specimens known to date. For more than 200 years, collectors considered the rarest and most expensive shell in the world to be the aptly named "Glory of the Seas", *Conus gloriamaris* Chemnitz, — a regal shell of the most elegant and fine colour patterns, which have been likened to the finest needlework. The Glory of the Seas cone has a legendary place in the history of shell collecting, and even though it is considered pretty and glossy but by no means spectacular, it is highly sought after.

Before 1837 just six specimens were known to exist, and even as late as 1968 only 40 had been found. From 1968 onwards, about 30 have been collected each year. In 1837, the first year of Queen Victoria's reign, the famous British collector Hugh Cuming made an extraordinary find — two individual specimens side by side under a small rock in the Philippines. He later wrote that he had nearly fainted with delight and amazement. The reef disappeared during an earthquake and it was thought that the only habitat of this rare cone had been lost forever.

The Glory of the Seas has been the subject of Victorian novels, and as recently as 1951 the rarity of this shell drove a thief (never caught) to break into a display case at the American Museum of Natural History in New York to steal a perfect specimen. This shell is still rather rare today, but a decent specimen can be bought for a few hundred Australian dollars.

Most of the great voyages of exploration during the eighteenth and nineteenth centuries were so-called "flag-planting" forays for new sources of minerals, spices, rare woods and medicinal herbs. But many resulted in very important shell collections and lavishly produced volumes, some with beautifully hand-coloured copper engravings that only the rich could afford (even by today's standards).

The size of shell books at that time ranged from tiny to absolutely massive and were quite cumbersome to consult. One extremely rare work (c.1824–40) on invertebrates, including many plates on shells, by Georg August Goldfuss, contains plates that are c. 43 x 60 cm. Some are double-paged and breathtaking in scope and detail.

Captain Cook's three voyages, undertaken between 1768 and 1779, made major inroads into knowledge of natural history and resulted in impressive collections. Other nations followed suit, the first large scientific expedition to the New World being financed and organised by the great German naturalist Alexander von Humbolt in 1799. Humbolt and others spent five lucrative years canvassing the as yet unexplored regions of northern South America.

In 1826, several years after Mawe's *Shell Collector's Pilot* had been published, France's Alcide d'Orbigny organised an expedition by the Paris Museum to southern South America, which resulted in a major contribution to knowledge of fossils and recent shells in this virtually unknown expanse. The resulting monograph on shells, *Voyage dans l'Amérique méridionale. Mollusques*, is a groundbreaking work on the natural history of the Southern Hemisphere. In his short life of 55 years, d'Orbigny produced a mind-boggling number of large tomes on natural history. His laboratory and personal library is still preserved intact at the Paris Museum, and visitors may use them for research purposes. Interestingly, many of the drawers in d'Orbigny's lab seem to have been untouched since he placed the specimens in their trays in the mid-nineteenth century. I found myself with black hands after inspecting fossil shells collected in the 1840s by d'Orbigny himself, and it was a major job just to clean off the tags in the boxes to see what was written on them, let alone the specimens themselves.

Also in the Southern Hemisphere, the globetrotter team of Quoy and Gaimard collected many unusual molluscs during their travels in the *Astrolabe* between 1826 and 1829. And again, very large format books were produced from the voyages, including the 1832–35 work *Voyages de découvertes de la'Astrolabe … sous le commandement de M.J. Dumont d'Urville. Zoologie: Mollusca*. These collectors were among the first in conchological history to publish exquisite, incredibly detailed, coloured illustrations of living molluscs, together with images of their internal anatomy.

Queen Victoria's reign saw an upsurge in interest in Mollusca, which has persisted to the present day, alongside a huge increase in knowledge about them. When Mawe's *Shell Collector's Pilot* was published in 1821, the entire list of all known species would fit on a few pages of text, and in 2003, at some 100,000 or more recorded species, it would be the subject of many volumes. This fact alone testifies to the importance and increasing popularity of shells through time.

THE LIFE OF JOHN MAWE (1766–1829)

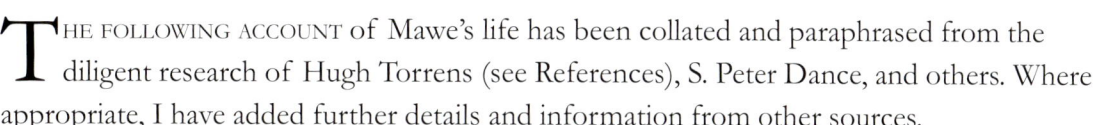

THE FOLLOWING ACCOUNT of Mawe's life has been collated and paraphrased from the diligent research of Hugh Torrens (see References), S. Peter Dance, and others. Where appropriate, I have added further details and information from other sources.

John Mawe's life was exciting and extraordinary — a perpetual treasure hunt! His desire to seek out the choicest shells and minerals, at almost any cost, and to sell them in his various shops around Georgian England, drove him on a daring quest around the globe.

Admiration for his tenacity and sheer determination is heightened when we learn of the extreme hardships of travel during the eighteenth and early nineteenth centuries.

And yet John Mawe has received very little attention over the years. Historically, it has been the likes of Linnaeus, Hutton, Cuvier and Darwin who have gained recognition as major contributors to our knowledge of natural history. But how did these scientists arrive at what were, for their times, astonishing conclusions about how the world and biosphere *really* work?

History has shown time and time again that those early naturalists who did not publish books or articles in academic journals have long been forgotten by the scientific world — many known only by their popular works. But it was these lesser known figures who advanced science in an important way.

The late 1700s and early 1800s was a critical period of enlightenment in scientific thought, with new ideas being discussed and old notions being challenged. For example, there was the question of whether a shell preserved in stone high on a mountain top was related to the 'Great Flood' or some other extraordinary process.

Naturalists and dealers such as Mawe were a special breed, who provided an immense wealth of new specimens and information for scientists, writers and collectors. The vast collections of shells that they brought back from faraway places were essential material for scientists documenting the composition of the land and marine fauna.

Mawe was undoubtedly motivated as much by money as by his passion for shells. In his books, he stresses time and time again that those who collected shells, packed them well, according to his carefully laid out instructions, and sent them to him in England, would be paid handsomely for their efforts. Nevertheless, he contributed a great deal to knowledge, not only by collecting huge numbers of shells, but also by publishing many books, including the *Shell Collector's Pilot*, the world's first shell collecting guide.

But I am jumping ahead. To start at the beginning, John Mawe was born in 1766 (not 1764, as almost every previous biography states), "in the house [in Queen Street] in which Mr Shaw the baker lives facing All Saints Church [now Derby Cathedral]". John's father was Samuel Maw(e) (1735–83), a baker and flour merchant in Derby. Sadly, no portraits of his son have come to light, despite an intense search.

It has been said that John was born of a very prosperous, middle-class stock. His mother, Elisabeth Massey, of Leeds in Yorkshire, became Samuel's first wife on 25 May 1759. The marriage was witnessed by Richard Brown (1736–1816), sculptor, marble worker and mineral dealer, who was to play a crucial role in John Mawe's career change from mariner to mineralogist and shell dealer in 1793. The family also had a connection with John Whitehurst (1713–88), the pioneer of Derbyshire geology, and Mawe later built on Whitehurst's original work.

John Mawe's mother died in 1776 and his father in 1783, so the poor young lad was orphaned at manhood's door — the tender age of 16. Samuel's second marriage to Fawnia Beighton had set the family up financially, as the Beightons had property. John's eldest brother, Samuel Sr., inherited most of the property and business, but luckily John received real estate in Derby and nearby Stanley. As a minor, he was also left in guardianship until

1787 to Joseph Sowter, Samuel's colleague and fellow baker at nearby Castle Donington in Leicestershire.

We do not know whether John spent much or any time with Sowter. What we do know is that his early years were spent at sea, as a mariner, in one of the merchant ships carrying Britain's expanding empire and trade around the world. But no details of his early travels exist, apart from a short statement in the the *Shell Collector's Pilot* (1821, p.23) that the ship on which he was an officer once floundered on an extensive coral reef in the channel of Mozambique, "from whence, and nearer the island, he procured some fine specimens".

Meagre information indeed, but it is certain that this early career as a mariner was crucial to John Mawe's later development as a natural history dealer who delighted the public and scientists alike with curious specimens of nature.

This was the time when "cabinets of curiosity" were still in fashion, and the collecting public would pay fortunes for choice specimens. The need to have the best collections drove the price of natural history objects very high and kept dealers like Mawe in a lucrative trade for many years. Mawe definitely had what it took to make a great deal of money from his travels and connections in the slowly growing world market, which he smartly developed over time.

We know that Mawe collected from "most parts of the globe", but it is uncertain whether some of these areas were visited by "scouts" paid to acquire specimens on his behalf. He either visited or collected in South America (Argentina, Uruguay, Brazil), Morocco, Mozambique, Comoro Islands in the Indian Ocean (between Madagascar and East Africa), Jamaica, and India. Of these, it is almost certain that Mawe personally visited Mogador in Morocco, Jamaica, Bombay, Tellichery, Brazil, and many islands in the Indian Ocean. In the *Shell Collector's Pilot* (1821, p.vi), though, he refers to the life of a mariner as a difficult one. Eventually, he decided to focus on selling natural history objects, and set up agencies to deal with acquiring specimens from individuals or at auction:

> Finding a sea-faring life hazardous during the war, after a favourable voyage I left off going to sea, and, with other business, commenced collecting minerals and shells: since which time, I may venture to assert, that the finest shells in the modern cabinets of Europe have passed through my hands, and to such an amount as would appear incredible to any one not interested in the science of Conchology.

WODARCH'S
INTRODUCTION
TO
THE STUDY OF CONCHOLOGY:
DESCRIBING
The Orders, Genera, and Species
OF
SHELLS:
WITH
OBSERVATIONS ON THE NATURE AND PROPERTIES
OF THE ANIMALS;
AND
DIRECTIONS FOR COLLECTING, PRESERVING, AND
CLEANING SHELLS.

A NEW EDITION,

With considerable Additions and Alterations,

BY

J. MAWE,

Author of Familiar Lessons on Mineralogy and Geology, Treatise on Diamonds and Precious Stones; Travels in South America, and through the Gold and Diamond Districts of Brazil, &c. &c.

LONDON:
PUBLISHED BY LONGMAN, REES, ORME, BROWN, AND GREEN,
PATERNOSTER-ROW;
AND BY J. MAWE, N°. 149, STRAND.
1827.

A popular treatise on shells by Mawe that went through many editions from 1822 to 1832. The first edition was by Charles Wodarch in 1820. The hand-coloured plates were finely executed by Edmund Crouch.

The war he refers to is, of course, the long conflict with France from 1793 to 1815, known as the "Great War" to Europeans until the close of the nineteenth century.

This suggests that Mawe must have been a Merchant Navy sailor until at least February 1793, when war was declared by France against Great Britain and Holland. Since he was a mariner for 15 or 16 years (both numbers being given in his works), he must have gone to sea in 1777 after the death of his mother, or in 1778 after his father's remarriage. He would then have been either 11 or 12 years old, which was not an unusual age at that time for enlistment in the British Navy.

After returning to dry land, Mawe joined the marble works in Derby, owned by Richard Brown II. The objective was to use a variety of Derbyshire fluorspars and marbles, via water power, to manufacture turned ornaments, which are highly collectable today on the worldwide antiques market. This move was good for Mawe's later career. Records reveal that Richard Brown was doing some wheeling and dealing in the mineralogical productions of Derbyshire, which no doubt influenced Mawe to pursue a career in capitalism. In 1784 Brown was elected as an Honorary Member of the London Society for Promoting Natural History. Patrons of the Derby Marble Works included influential figures such as Erasmus Darwin (Charles Darwin's grandfather), Josiah Wedgwood of "porcelain" fame, and geologist Faujas de St Fond.

While there, Mawe became friendly with the boss's eldest daughter, Sarah, and married her on 1 November 1794. As far as we know, the couple had only one child, a son named John Saint Mawe, born on 21 February 1798, who was reputedly brilliant and won prizes at university. Sadly he died at the age of 22.

Shortly after his marriage, Mawe moved to London to take up a manager's position in a new shop, together with Richard Brown II and III, called Brown, Son and Mawe, Petrifaction Warehouse, at 5 Tavistock Street, near Covent Garden Market. By 1797 there were two distinct aspects to this business – one was the sale of minerals and the other, the "Brown and Co." part, selling the spar and marble ornaments made at their jointly managed and owned Derby Marble Works.

The business had very much expanded by this time and helped Mawe to find his way around the mineral market. He soon was to make a name for himself in scientific circles, and was flattered when politicians sought his scientific opinion. This led to Mawe preparing for Lord Pelham an outline of a proposed mineralogical expedition to New South Wales and Ceylon in 1799/1800. But the voyage apparently never took place.

In 1802 — an important milestone in Mawe's career — Brown, Son and Mawe moved their establishment to a new factory in St Helens, Derbyshire, where they installed a six-horsepower steam engine. He also published his first book of many, *The Mineralogy of Derbyshire with a Description of the Most Interesting Mines in the North of England, in Scotland, and in Wales*. In it he writes about spending a great deal of time in the mines near Castleton, which he calls "the most interesting part of the county", and where he "first became attached to his favourite study of the science of mineralogy".

Also in 1802, a peace treaty was signed between France and England, called the "Peace of Amiens", which meant that there was much more freedom to travel between these powerful nations. Mawe went to France during the latter part of the year to deal in minerals and to visit the most renowned geologists of the time, including Hauy and Faujas. Mawe was not enthused at all about learning French, lamenting that he "would certainly rather go on an East India voyage as a sailor (and it would be less difficult) than to learn French." But fortunately Faujas spoke English, and after several months in the country, Mawe's French is said to have improved a little.

Mineralogists of the time declared that Mawe rivalled the best dealers in the business, which attests to his proactive, capitalist nature. In 1804 he issued a catalogue, which claims that "collectors will find the largest Variety of Minerals and Shells in Europe" at Mawe's Covent Garden shop. In his 1804 treatise on natural history, the first edition of what would later be the *Shell Collector's Pilot*, he writes on the title page that his establishment has the "largest variety of natural history in Europe" — a bold advertisement but apparently true. Mawe had arranged for the supply of specimens from Sweden, Norway, Switzerland, Madrid, Italy, Paris, East Indies, New Holland, ... The world was seemingly his oyster as he strove for more and more choice specimens.

Content that his business was booming, on 1 August 1804 Mawe set off on his long "voyage of commercial experiment" to South America. As he headed for Rio de La Plata, war was declared between England and Spain, and he was detained in Cadiz. He was not able to reach Montevideo in what is now Uruguay until 1805, where he was restrained as a British spy and imprisoned until 1806, when Montevideo was captured by the British. He then made his way to Brazil via Buenos Aires under the protection of General Whitelocke, but eventually purchased his own schooner. Stopping at the southern ports of Brazil, he finally arrived at Rio de Janeiro.

From here, he made his famous journey to the Mineas Geraes. His status as a qualified expert on natural history, with Royal (now Portuguese) patronage, guaranteed his access to mining sites not otherwise easily visited by a European in Brazil. The success of his book *Travels in Brazil* undoubtedly stems from this first authentic, witnessed account of the country's diamond and gold mines. The view of the Jaragua mining camp is the first published landscape image of the interior of São Paulo.

First published in 1812, *Travels in Brazil* is valued by geographers and historians as one of the earliest reliable descriptions of Brazil under Portuguese colonisation, and of the diamond and gold washings. It is also replete with details on the country's geography, anthropology, fauna and flora, and other information. As a result, Mawe has been described as a "shrewd and intelligent observer".

In 1813, he published *A Treatise on Diamonds and Precious Stones*, which, along with *Travels*, was translated into many languages and editions. In the *Treatise* (p.168) he underscores his own abilities to obtain the best specimens available on the market: "Mr MAWE flatters himself that he has been able to form connexions [sic] by means of which the finest productions of those rich countries will be transmitted to him; the whole of which will be sold on the most reasonable terms."

A year after returning to England late in 1810, he opened a shop at 149 The Strand, London, where he not only entertained students and gave demonstrations but most importantly sold natural history objects and published several major books. Making the most of his status of Royal patronage, he dedicated *Travels in Brazil* and *Treatise on Diamonds* to the Prince Regent of Portugal, later King John VI of Portugal, claiming for himself the title of "First Administrator and Mineralogist". His 1818 his *Catalogue of Minerals* recorded his membership of the Royal Geological Society in Cornwall.

It seems that these claims to patronage opened the way to success as a writer and publisher of popular textbooks on mineralogy, shell collecting and shells, and as a supplier of rare, precious specimens as a "Royal" museum proprietor in London. In 1823, a few years after the publication of *Shell Collector's Pilot*, Mawe wrote one of his most popular and important shell books, *The Linnaean System of Conchology, etc.* He also published several editions of the popular *Wodarch's Introduction to the Study of Conchology: Describing the Orders, Genera, and Species of Shells*....

In a very rare four-page advertisement, entitled *Works on Mineralogy and Conchology*, published by Longman in c.1827–28, Mawe lists no fewer than 11 books that he had published in various editions. His popular work *Lessons on Mineralogy and Geology* was already in its ninth edition. Another advertisement, in one of his books on mineralogy, provides Mawe's estimate of the value of shells at that time, 1824 (the price increasing to five guineas for collections by 1828):

> SHELLS – The finest assortment of rare and valuable Shell, price marked on each.
> *Collections* shewing the genera, from Three Guineas to Ten or Twenty, according to the number and quality.
> More extensive Collections of Shells, from Fifty to One Hundred Guineas and upwards.

John Mawe's wife, Sarah, had run his business most successfully while he was away from 1804 to 1810, and clearly acquired his ability to attract royal patronage. Well after her husband's death in 1829, she was appointed "Mineralogist to her Majesty" Queen Victoria, just after the coronation in 1837. Sarah's contributions were recognised by mollusc researchers such as early Victorian scientists Griffith and Pidgeon, who named the beautiful muricid, coral-dwelling gastropod *Latiaxis mawae* (today living in Japan, Taiwan and the Philippines) in her honour in 1834.

Further books by Mawe were published posthumously. The fourth edition of *Wodarch's Introduction to the Study of Conchology* was published in 1832, three years after his death, no doubt by Sarah Mawe. This late edition included newly discovered species and amendments in the allocation of species to various genera, and was more than likely in the last stages of preparation by the time of Mawe's death.

By the time he died at the age of 63, John Mawe had seen more of the planet than most Europeans could begin to imagine. This great man, who appreciated and cherished the jewels of the natural world, should be admired by scientists for his significant contributions to natural history, and by capitalists for forging a career in gaining wealth from organic and inorganic beauty from the far reaches of the globe. Mawe appropriately ends his treatise on shell collecting with a piece of advice:

> PERMIT me to advise the traveller to look into the book of nature, which is always open, and learn what he can...

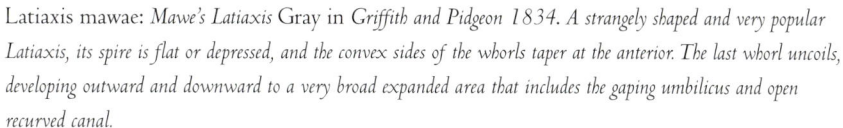

Latiaxis mawae: *Mawe's Latiaxis* Gray in *Griffith and Pidgeon 1834*. *A strangely shaped and very popular Latiaxis, its spire is flat or depressed, and the convex sides of the whorls taper at the anterior. The last whorl uncoils, developing outward and downward to a very broad expanded area that includes the gaping umbilicus and open recurved canal.*

FACSIMILE OF THE FIRST EDITION 1804

John Mawe

1804

A SHORT
TREATISE,
ADDRESSED TO
Gentlemen visiting the South Seas,
AND ALL
FOREIGN COUNTRIES;
MORE PARTICULARLY TO
Commanders, &c. of Ships,
AND
GENTLEMEN RESIDING ON SHORE,
With a View to encourage the collecting of
NATURAL HISTORY.

By JOHN MAWE,
Author of the MINERALOGY of DERBYSHIRE, &c.
No. 5, Tavistock Street, Covent Garden, London,
Where Collectors will find the largest Variety of
NATURAL HISTORY in EUROPE.

YOUNG, Printer, Brydges Street, Covent Garden.

1804.

ADVERTISEMENT.

Mr. MAWE begs to inform the Amateurs in Mineralogy, that his Collection generally consists of all the known Substances, and from his having formed Connections in *Petersburgh, Sweden, Copenhagen, Norway, Vienna, Switzerland, Madrid, Italy, Paris, East Indies, New Holland,* &c. &c. it may be presumed he is likely to meet with every new Substance as soon as discovered, which Collectors may purchase at the most moderate Prices. His Cabinets exhibit above 20,000 Varieties, which are always open for Public Inspection.

Small Collections described, consisting of Forty or Fifty Specimens, Two Guineas.
Ditto, ditto, consisting of One Hundred Varieties, Five Guineas.
Collections consisting of Two Hundred Varieties, Ten Guineas.
Grand Collections, from a Hundred to a Thousand Guineas.
Chemical Tests and Mineralogical Chests.
Sets of the primitive Form of Crystals.
Mineral Substances and Earths for Chemical Experiments.
A large Variety of Spars, Shells, &c. for Grottos.

No. 5, *Tavistock Street, Covent Garden.*

A
SHORT TREATISE
ON
NATURAL HISTORY.

THE Knowledge of Natural History is a truly scientific Pursuit, and can only be attained by great Study and Practice. It is not requisite that a Collector should be completely Master of the Science; but by preserving all the Varieties that he may meet with, would be advancing himself in the Pursuit.

SHELLS form a considerable Class of Natural History, and are important to the Scientific. The most beautiful Dye has been extracted from the Purpura, and the Animals have, in many Instances, been found of the greatest service for Ailment in pulmonary Complaints. Exclusive of the peculiar and learned Science, which respects the form, Mark, and Species, and determines to what Class they belong, this Study informs us in our Cabinets of the various Inhabitants of the Sea in distant Climates, and will amply repay those who take the Pains to collect them.

Land

LAND SHELLS are found in damp and swampy Places, frequently upon Trees, and almost in every Ditch or hollow. They are generally more tender than Sea Shells, and are equally esteemed. It is necessary to observe, that Shells are in general tender, and the smallest Damage, as a Snip or Bit broken from them, renders the Shells (almost) good for nothing; therefore too much Care cannot be taken of them.

If a Shell lies in the Heat of the Sun for a few Hours in a hot Climate, after the Animal is taken out or dead, it is injurious to it; therefore they should be kept out of the Sun. Shells gathered from the Beach are generally bleached by the Heat, and as frequently broken by Birds picking out the Fish. The best Shells are found immediately after the Tide leaves them, or in shallow Water, before the Birds have seen them, or are brought out of deep Water by a Dredge, or by People diving for them, and frequently by hauling the Siene. After Gales of Wind valuable Shells are frequently washed into shallow Water, or left on Shore,

In order to get the Animal out of the Shell, put it in Water nearly boiling hot for two or three Minutes; then take it out, and in a Minute or two after put it in cold Water, where let it remain ten Minutes; the Animal then becomes more solid, shrinks, and will generally come easily out. A hot Iron is sometimes applied to the Shell, to force the Animal out: this Method should never be practised, as it invariably spoils the Shell.

If there is a hard Substance that shuts the Animal up in the Shell as its Lid or Operculum, it should be carefully preserved.

If the Shells have a disagreeable Smell, they should be put into a Box, which should be filled with Saw-dust, or dry Sand, and nailed down.

Shells are no worse for being dirty; they had better be brought with all their native Characters, for washing and scrubbing frequently spoil them.

One Remark, not the least worthy of Notice, is, that valuable Shells are frequently spoiled after they are collected, by being put in empty Beef Tierces, &c. without packing, where they rub one against another as the Ship rolls, by which means their Characters are entirely defaced and spoiled. It cannot be too strongly recommended to pack them in a Box, and nail them down as soon as they are collected.

Every

(6)

Every Rock affords Limpets. They should be taken off in such a Manner as not to injure their Edges; they are curious and interesting, especially from new Countries.

Shells found on old Wreck, or Wood that has long lain in the Sea, should be preserved; also Barnacles, if large.

Shells found in Mud should be preserved: it is not pretty Shells that are alone interesting; they are in general the most common; it is new Varieties that the Scientific want; but by preserving *all* that may be found, is the certain way of doing right.

A Shell-like Substance in straight Tubes and round Whirls, or Coils, is frequently found in Sand; they should be preserved with Care, as they are generally very tender.

Rocky Coasts and narrow Streights in general produce the best Sea Shells; and swampy Lands, stagnated Waters, Ditches, &c. the best Land Shells.

If at any time Snails are exposed in the Market of any Country for Food (with their Shells) as is common in *France*, *Spain*, and *Italy*, it is particularly recommended to buy a Dozen or two of *every* Variety; this may, if attended to, be the best Means of discovering

(7)

covering Aliment of the greatest Service in Consumptive Cases, or Diseases of the Lungs.

Shells adhering to the Branches of Trees, as the Mangrove and others, a small Branch containing them should be cut off.

All kinds of Fresh-water Shells, as Muscles, &c. should be preserved; also Oysters, Cockles, Whelks, and others, of every Description. It is worthy of Remark, that the most common Shells in the South Seas, and many other Countries, may be very scarce, or perhaps not known in *England*.

Shells or Barnacles adhering to the Whale, Shark, or other Fish, are very interesting, and should be preserved.

Wherever there is a Fishery for Oysters or Pearls, the Divers frequently bring up many rare and good Shells; Varieties of all should be selected and attended to.

The Author will very gladly supply any Gentleman with Boxes, &c. gratis, and will with great Pleasure pay any Expence, or purchase any Collections of Shells or other Substances, that may be made during a Voyage.

(8)

MINERALOGY.

THIS useful and interesting Science, upon which depends the Riches, Commerce, and Wealth of (perhaps) all civilized Nations, is now generally studied, and forms a Part of almost every Gentleman's Education. To know how Nature has stored the Earth with the immense Variety of Riches, so serviceable to the Purposes of the Human Race, is perhaps presumptive, and beyond our Comprehension.

Of the component Parts which form the Strata of *Europe*, we have some slight Knowledge; their Productions are converted to Purposes necessary to our very Existence.

The Grandeur of *England* may most unquestionably be dated from her Coal Mines, they form the Basis of all her Manufactures and Commerce.

Russia and *Sweden* may date their Commencement as Nations from their Iron Mines, Copper Mines, &c. *Norway* from her Silver Mines, *Portugal* and *Spain* from their Mines of Diamonds, Gold, Silver, &c. Many more Instances might be brought forward were it necessary, but this is only meant to

(9)

stimulate the Mind to pay Attention to the various Metals, Rocks, &c. that may be met with in remote Countries; for it should be remembered, that the roughest Rocks produce the most valuable Substances, and the most valuable Diamonds and Metals have been accidentally found in little Water-courses, and Chinks of Rocks.

It is much to be wished that we had a Specimen of every different Rock found in distant Countries; by that we should know whether it was Granite, Limestone, or any other Substance; and might in many Cases judge, by comparing it with others, what Minerals would be probably found in it. Coal and Limestone, lately discovered in *New Holland*, are likely to prove much more valuable to that fifth Division of the Globe, than Mines of Gold or Diamonds. Specimens, as a Piece of Stone, no matter its Appearance (they need not be pretty to render them interesting, our Coal and Iron Ores are not pretty) should be broken from the Rock, and not picked up from the Beach, as the Appearance of the Fracture forms an important Character, which would be destroyed were it rounded or mutilated by Attrition: these should be wrapped up and packed with a Label to each, specifying where it came from. Would Gentlemen who frequent new Countries be at the pains to collect every Variety of

Ores

(10)

Ores used by the Natives, or Stones, Rocks, Clays, &c. used for Building or any other œconomical Purpose, or break a Piece (three or four Inches square is sufficient) from any Rock or Island they might have Occasion to land upon, they would render *Natural* Geography the greatest Service; from them we should form some Idea of the component Parts of the Island, &c. they were broken from, and the Geology of the Southern Hemisphere would soon be known. The Advantages arising therefrom are impossible to be foreseen, as from Specimens so collected, the greatest Benefit might accrue to our Commerce.

Gentlemen making such Collections would be doing an essential Service to their Country and themselves, as Analises and Examinations would immediately take Place, and the Collector be amply rewarded according to his Merit.

This is most certainly the most useful and interesting Branch of Natural History, and cannot be too much encouraged, and in which we find Men of the highest Rank engaged for the Benefit of Mankind in general. It need only be asked, how many Obligations are we under to the Discoverer of Iron, &c. &c. &c.

New Metals or new Earths, or Varieties, may prove of the greatest Service.

(11)

As these are the two most important Branches of Natural History, I have treated of them more at large than I meant. Vegetable Productions are of great Importance, but require great Care in their Preservation, as do Insects and all the animated Creation; therefore I shall only say it is much to be recommended, that all should be preserved by those Gentlemen who have Zeal and Taste for the Pursuit.

Any Information may be obtained relative to Natural History, and every Instruction given, with Packing Boxes, &c. requisite for their Preservation, by applying to Mr. Mawe, No. 5, *Tavistock Street, Covent Garden.*

Gentlemen desirous to dispose of their Collections of Natural History, may consign them to Mr. Mawe, who will pay the most liberal Ready-money Prices for them, or dispose of them on Account, and give Security to any Amount. Reference to Mess. Ransom, Morland, and Co. *Bankers, Pall Mall.*

Gentlemen who will take the Trouble to send a Box from any Settlement, &c. may depend on the most liberal Returns.

FINIS.

FACSIMILE OF THE COMPLETE 1821

The Voyager's Companion; or Shell Collector's Pilot

John Mawe

THE
VOYAGER'S COMPANION;
OR
SHELL COLLECTOR'S PILOT:
WITH
*Instructions and Directions where to find
the finest Shells;*
ALSO
FOR PRESERVING THE SKINS OF ANIMALS;
AND THE
*Best Methods of Catching and Preserving Insects,
&c. &c. &c.*

BY JOHN MAWE.
Author of Treatise on Shells, Lessons on Mineralogy,
Travels in Brazil, &c.

Third Edition, with colored Plates.

LONDON:
PRINTED FOR AND SOLD BY THE AUTHOR, 149, STRAND;
AND LONGMAN, HURST, REES, ORME, AND BROWN,
PATERNOSTER-ROW.
1821.
Price Five Shillings.

ADVERTISEMENT.

MANY have been deterred from sending shells, fearing they might be seized by the officers of the customs. The following is a certain mode to prevent it: and it is particularly recommended to captains of ships, passengers, &c.

Any person desirous to send a box of shells &c. to the author, he will pay the charges, and make an adequate return in whatever way the consignor may direct. It is proper to observe, that they must be packed in cases, and entered on the ship's manifest, and marked as below, which will prevent any risk of seizure.

J. MAWE,
King's Warehouse,
LONDON.

N. B. Ships bound to any out-port, if the case be directed as above, it will be duly forwarded to the address.

DESCRIPTION OF PLATES.

The Frontispiece.

Low water, and the beach strewed with most beautiful shells and coral, to tempt the traveller, and rouse his mind to contemplate on the beauties of the deep, and the wonderful works of an Omniponent Being—

"Who taught the little nautilus to sail,
"Spread his thin oar, and scud before the gale!"

The bird is emblematical of—

"Search, and ye shall find."

A few snails are placed on the land, shewing, that although the animal is not pretty, yet he forms a shell often held in great estimation.

A few tropical birds are introduced, to show another variety of the beauties of the creation.

The plate of *Insects* requires no explanation.

INTRODUCTION.

IN offering this little Work to the Public, more especially to sailors and travellers, it may not be improper to state, that the following pages are generally the result of my own observations, during fifteen years that I was at sea, and subsequently whilst I was resident for six years in South America, and the Brazils.

INTRODUCTION.

HAVING sailed to most parts of the globe, I may say, from my own experience, that there is no station which affords such facilities for collecting shells and various subjects belonging to Natural History, as that of commander or officer of a ship, whether he please to make it an amusement, or a traffic.

FINDING a sea-faring life hazardous during the war, after a favorable voyage I left off going to sea, and, with other business, commenced collecting minerals and shells: since which time, I may venture to assert, that the finest shells in the modern cabinets of Europe have passed through my hands, and to such an amount as would appear incredible to any one not interested in the science of Conchology.

IN this, as well as in every other branch of natural history, the knowledge requisite to discover the scarce and rare varieties from the common, can only be acquired by practice. To particularize objects that may present themselves to the traveller, is not the aim of this work; but, for the benefit of science, and his own interest, it is desirable to excite him to collect all he may meet with, until he has skill to select the good from the bad;

and I can assure him that he will be amply remunerated for his labor. I say *all*, for however common they may be in the countries he visits, they may be scarce here.

Several years ago, I published a small pamphlet, entitled " Directions to Captains of Ships, Officers, and Travellers, particularly to those who visit the South Sea Islands," &c. which went through two editions. Since that time, science has rapidly advanced, and it may be said, there are but few who have not felt a desire to obtain something more than an ordinary knowledge of the productions of nature. The accomplishing of this has been rendered extremely easy by recent publications, which embrace, separately, the various branches of natural history, more particularly that of Shells*, describing and explaining the various orders and species, with particular instructions to collectors.

When at sea, I have often lowered down my boat to take in floating wood,

---

* The Author has just published a small Treatise on Shells, with colored plates, and lists of the names of the various species under each genus, written expressly for learners.

x INTRODUCTION.

(wreck), sea-weed, &c. from which I have collected many rare small shells.—Whales have frequently barnacles adhering to their heads and jaws; and the wood is often penetrated by the destructive worm, (teredo navalis)*; which is extremely interesting to examine.

---

* The Teredo Navalis is a mucilaginous substance, in the head of which are strong muscles, terminated with hard shell-like substances, one of which is not unlike the cutting part of a carpenter's auger: this the animal works so as to bore holes in almost every sort of wood. It is supposed that the animal, when extremely

INTRODUCTION. xi

LIMPETS may be found in every harbour, on every rock, and on every coast: the sea seldom ebbs without leaving shells of various species; among the most common that are exposed for sale in the markets of the countries the traveller may visit, the connoisseur might discover some rare or interesting varieties.

I SHALL conclude my observations up-

---

small, floats in the water, and attaches itself to the bottoms of ships, (if of wood), which it almost immediately pierces, and, like the pholas, grows and becomes larger as it penetrates.

on this subject, with strongly recommending the reader, whenever opportunity occurs, as ships loading, refreshing, &c. to employ the fishermen on the coast to collect for him; these men are well acquainted with the places where shells may be found, and for a trifling remuneration would gather a supply, which, on his return home, might gratify his friends, or otherwise be turned to advantage.

It would also amply repay him, to hire an expert negro to go into the interior in quest of LAND-SHELLS; for, although the animal be a snail, with his house on his back, and less beautiful than sea shells, yet they are interesting and desirable, from being seldom attended to. In Brazil, at the royal farm at Santa Cruz, where I resided some months, holding a high official situation, I adopted what I here recommend, and succeeded to the utmost of my wishes.

The following is the Recipe for making the *Preservative Soap*. As both it and the *Powder* are deadly poisons, I have thought proper, by adopting the technical phraseology, to conceal (in some degree) the ingredients from those who might apply them to improper purposes.

Arsenici Oxydi .... ℥j
Saponis .... ℥j
Potassæ Carbonatis .... ʒvj
Aquæ saturatæ .... ʒvj
Camphoræ .... ʒij

Preservative Powder.
Arsenici Oxydi pulvis.

CONTENTS.

Introduction v
Chap. I. *Cleaning and packing shells* 1
II. *Localities of shells, &c.* .. 5
    Ambergris 26
III. *On Insects* 29
IV. *On Birds* 35
V. *On Reptiles.* 39
VI. *On Quadrupeds* 43
VII. *On Plants* 49
VIII. *On Minerals* 53

CHAPTER I.

On Cleaning and Packing Shells.

SHELLS frequently receive considerable injury, and sometimes are entirely spoiled, by the attempts of unskilful persons to clean them.—— It is therefore of essential consequence that the following observations be most strictly attended to.

When a live shell is found, it would soon become offensive, unless the animal were taken out. To perform this, nothing more is requisite than to put the shell into a kettle of water, and let it heat gradually, until it boils.

B

After a few minutes, the shell should be taken out, and put into a bucket of cold water; the animal will then shrink, and may generally be shaken from the shell: but if it should still adhere, it may be pulled out with a crooked pin or hook, great care being taken not to injure the mouth, which is commonly the most tender part of the shell.

Oysters, muscles, clams, or limpets, may be treated in the same way; or they may be opened with a knife, and the animal cut out, which must be done very gently, least the shell should be chipped or broken: but the former method is preferable, since the shell opens of itself, when in boiling water. The same care should be taken in separating limpets from the rocks, for the least chipping renders them but little esteemed.

This operation being performed, the shells should be immediately rinsed and stowed away; and no further attempts at cleaning or polishing should be made. The rough outside, (epidermis), forms a principal character in the shell, and should on no account be removed——the perfection of a shell depends on its being in its *natural* state.

Many shells are in themselves such beautiful objects, that the traveller should be as expeditious as possible in removing them out of sight, for they invariably create in the beholder a desire of handling them, which is generally injurious to the interests of the possessor.

We shall now proceed to describe the best methods of *packing* shells. Pour some sawdust, or, if that be not at hand, some sand from the beach, into a chest or beef barrel; into this the large strong shells may be put, covering them with sand, or sawdust. For the more tender varieties, small boxes, about a foot square, and six inches deep, should be provided, (which may be purchased for a shilling each), into which the shells should be placed in layers, with sawdust strewed amongst them. The Paper Nautili should be packed, keel down, in shallow boxes, which should be filled with sawdust, moss, cotton, or paper shavings, for any heavier substance would chip the edges of the shells, and diminish their value. The boxes into which these are packed, should never contain more than one layer. Small chip boxes may be used for particular varieties, but as these boxes are very slight, they should be afterwards packed within the larger ones; and when all are full, they ought to be stowed away in an empty barrel or chest, which should be finally closed or nailed down; marked, and entered on the *Ship's* manifest, as SHELLS—to prevent the risk of seizure. The Custom House expenses are so trivial, and the process so simple, that the Author will have pleasure in assisting any one who may be unacquainted with the routine.

CHAPTER II.

Localities of Shells, &c.

SHELL-FISH are generally esteemed for food, and a great variety of shells are in the market of almost every sea-port, both abroad and at home: at Billingsgate, Portsmouth, Plymouth, &c. we have our periwinkles or whelks, muscles, cockles, scallops, gapers, and oysters; the same species are also exposed for sale in the markets of other countries, but they differ from those which are found on the English coasts. The pectens, scallops, cockles, &c. from France and Spain differ from our's, which renders it desirable to possess them. The common shells (except the oyster), which are found at Cadiz and in the Mediterranean, especially at Cette, Marseilles, and Genoa, are desirable. When at Mogador, I found many good limpets, and scarcely any thing else worth notice, except some muscles and *land-shells**, which I persuaded the Moors to bring from the Interior. We receive many varieties of *land-shells*†, (some of which are large, and particularly in request), from the Gambia and the interior of Senegal. At Goree, and along the African coast many varieties of shells occur, as well as to the southward, at Sierra Leon, and more especially Cape Palmas, the Gold Coast, and Bight of Benin. From the islands, St. Thomas's, Annabona, and the coast about Loanga, very good shells have

---

* Snail shells.
† The animals constitute an article of food; and I have often seen snails, boiled in their shells, served up with rice, in various ports in the Mediterranean.

been brought; also from the vicinity of the Congo, and the Portuguese settlement, Benguela. One of the rarest shells known, and several scarce varieties, have been gathered from these coasts, also sea-fans (*gorgonia*), and interesting weeds; all of which, as well as the *land-shells* from the interior, are desirable.— Farther south, as far as Saldahna Bay, the shells are similar to those about the Cape.

ALONG the coast of North America, and as far as the Spanish Main, few interesting shells have appeared: from the latter numerous varieties of volutes and camp-shells, *(pen-a-mar)*, are brought continually. We know but little of the shells that may be found about Pensacola, or New Orleans, or along the coast until we arrive at Vera Cruz; but from this port, Carthagena, and the intermediate coast, we have received a few fine varieties. Some beautiful spined shells, of the Venus species, (not unlike cockles), have been brought from Trinidad and the shoals of the Oronoco.

THE West-Indies do not produce many of what are termed rare shells*. I have picked up fine conchs, sea-fans, and weed, all over the coast of Jamaica, and in the interior some interesting *land* and lagoon shells.

A FRIEND of mine, a commander of a ship, who went to Demerara, employed at my request an expert negro to go into the Interior to collect *land* and fresh water shells. The man was absent a week, and collected a box-full of what are termed *snail-shells*, and muscles; these were packed with refuse cotton, and the whole expense did not exceed six dollars. And let me here recommend the traveller to hire a fisherman, or clever negro, (who ought to be well remunerated on his return), to collect, as well

---

* The collections of shells, in fine boxes, which are sold in the West-Indies, are too contemptible to notice; they can only please mere novices.

as to pack these objects, and thus free the employer from any trouble.

THE *land shells* and muscles which may be found over the whole of the vast territories of Surinam and Cayenne are very desirable. I would amply repay any one for what he might bring from the interior of those countries.

OF the shells from the Amazons, the great island Joannes, and from the district of PARA, we know little or nothing, except that there are in the interior many fine *helices* (snail shells), muscles, &c.——Shells, therefore, however common in those places, would be objects of attention here.

FROM Pernambuco to Bahia and Rio de Janeiro, the *land* and fresh water shells are quite as interesting, if not more so, than those which are found on the coast; to the southward of

C

Rio*, near the isle of St. Sebastian, the Paper Nautilus, and other fine shells, are frequently met with. About six years ago, after a gale of wind, a spring tide ebb left a reef of Nautili and other shells along the coast of Bahia, mutilated by the surf of the conflicting elements: a Government-ship was at the same time loading with timber. On her return to Chatham, I received intelligence from an officer on board, that he had collected a quantity of them. I sent a person down, who purchased to the amount of twenty-five pounds. A similar circumstance occurred under my own observation, at the mouth of the Guadalquiver, near Cadiz, from whence I select-

---

* At the Royal Farm, Santa Cruz, about forty miles from Rio, where I held an official situation, (first administrator), I directed some expert negroes to pick up what snail shells and curious animals they might meet with: these they left at my house as they passed, and, by allowing them a small compensation, I obtained many fine shells, insects, birds, reptiles, and small animals of the monkey, ape, and hedge-hog species.

ed many fine varieties. Indeed, almost every gale of wind throws up some interesting shells.

From the interior, near Santa Cruz, also about Santos and Bertiojo, I collected many curious shells; but strange to tell, I have found it difficult to persuade any one to send me another supply, although so easily obtained. No, they think them common, and because they *really are so there*, they do not bring or send them.

When at the isle of St. Catherines, and the bays near it, I employed the fishermen to collect for me, and obtained from them many varieties of shells, sea-eggs, star-fish, coral, and sea-weed, also muscles and land shells from the lagoons, to which I gave the preference.

In the river Plata I was not equally fortunate; for though a gale, called a Pampero, had left the river dry in many places, from two to three miles in extent, I could discover no shells;

nor were the fishermen, whose attention I engaged, more successful. Notwithstanding, I am told, fine volutes are met with there.

Many fossil-shells may be found under a black soil, resting on granite, near Monte Video.

I collected some good shells in the interior, and at Barriga-negra, near the river Sebollitee, where I was detained many months*.

The rocks which form the Falkland Islands, produce very fine limpets. Many good shells have been brought from Magellan Straits, and Staten Land; nor are the remote islands, called South Georgia, or Kerguelen's Land, without fine limpets, and thick strong shells, which are desirable.

---

* See the Author's Travels through the gold and diamond district of Brazil.

After rounding Cape Horn, we know nothing of the shells that may be produced on the coast of Chili and the neighbouring islands; nor indeed, until we reach the coast of Peru, from whence many fine varieties have been procured, especially from the shores of Callao, and near Lima. All the species that may be found at these places, at low water, or obtained from the fishermen, should be carefully collected.

The Gallipagos Islands, we know, are rich in shells, and it is wonderful that more have not been brought from thence.

From the islands at the entrance of the gulf of California, and from the rocks and beach, numerous fine Ears have been collected, for which in one year I paid above a hundred pounds.— The barnacles which adhere to the whales in these seas, are different from those which are found on the other side of the continent; so are the limpets, clams, muscles, &c. consequently they are highly interesting.

In crossing the Pacific, the Sandwich Islands are sometimes visited; the shells from whence are in great request. To the south of the line, is the rich group of the *Marquesas* and the Society Islands, from whence we have *many* beautiful and rare shells, chiefly collected by Circumnavigators. Here the commanders of whale ships, &c. are particularly requested to pay every attention, and to gather up whatever is strewed on the beach, and the limpets, which adhere to the rocks, as well as what land and fresh water shells they can possibly procure.

The shells from Dusky Bay, New Zealand, and all along the coast, also the reefs about New Holland, particularly the western part, King George Island, the Marian Islands, Port Jackson, &c. will most amply remunerate the trouble of collecting: and let me here state, that for several years I paid more than two hundred *per annum* chiefly for shells, &c. to gentlemen holding the first situations* under Government. The finest lot of a peculiar species ever brought to this country, was gathered by two boys in Western Port.——A whaler off the coast sent a boat on shore to search for fresh provisions, as birds, animals, &c. whilst the crew were shooting, the boat grounded among stones and weeds, and during the time before she floated, the boys left in charge of her, employed themselves in gathering the shells entangled in the weeds and about the stones. On their return home they

---

* One of the finest collections of shells I ever bought, was from a Lieut. Governor of St. Helena, who was a connoisseur. He assured me that he obtained them from South Sea ships and Dutch, Danish, and Portuguese, Indiamen, whose commanders, wanting refreshments, found it their interest to present him with any fine shells they might possess.

brought them to me; and though I gave them the price they asked, I thought it not only right, but political, to present them with a guinea each, to stimulate them to look out for shells on another voyage. I am sorry here to add, that it is, generally, only the boys or the cook, who notice these *rarities*, and who make a few pounds by them every voyage.

From Tongataboo, one of the Friendly Islands, Bligh's Island*, and the cluster of Fejees, some varieties of extraordinary beauty have been brought.

---

* Named after its discoverer, Admiral Bligh. His lady possessed one of the finest collections of shells in Europe. The admiral having twice circumnavigated the globe, and being afterwards Governor of New South Wales, she was enabled to enrich her collection with the most rare and valuable species from all parts of the world. This extensive and fine collection is now in my possession.

From New Caledonia, and the vast group of the New Hebrides, we possess no shells whatever; but from the coasts of *Papua* and New Guinea, some very rare varieties have been received.

The shells from the Chinese seas are generally interesting; from the Philippine Islands we have many fine varieties: the Dutch, Portuguese, and Danes, have also contributed to our cabinets in this department. At *Wampoa*, or Canton, shells from Japan, Formosa, and Haynan, may be purchased in shops of little importance. If any of the commanders or officers of our China ships would take the trouble to employ some Chinese fishermen to collect land and fresh-water shells, he might, for three or four dollars, depend upon obtaining whatever the country produced: and were he to extend his order to ten dollars for sea-shells, he would be amply repaid on his return home. The very commonest productions in China have been ne-

D

glected; they may probably, from their constant occurrence or uninviting appearance, have not been thought worthy of notice: whilst on the other hand, carved nautili, and large green shells, which have been ground, (and therefore injured), to display their pearly lustre, have been seized with avidity, but have failed to repay the collector, or gratify the connoisseur.

GOLD and diamonds, (which are found in the soil of the rivers), have been brought from Borneo, but we are totally ignorant of what shells may be produced there.

FROM the cluster of the Celebes, we have a few fine varieties, which have been noticed by officers of ships of war, or circumnavigators; but, strange as it may appear, the Author of the Narrative on the Pellew Islands has not noticed or described one single shell from thence.—— One of the rarest and finest shells ever seen was brought up in the mud sticking to an Indiaman's anchor, when getting under weigh, in the straits of Macassar.

FROM the islands in the Archipelago between the north coast of New Holland and the continent, but more especially from TIMOR and Amboyna, many valuable shells, as well as beautiful corals, have been brought.

A MAYLAY fisherman was employed for a fortnight by a friend of mine, whilst at Timor, from which I reaped considerable profit.

FROM Java, Sumatra, and the Malay shore, many shells have been brought, but by far the finest were collected at Bencoolen, by a gentleman high in the civil service of the Honorable East India Company, who employed a fisherman at my request.

FROM the Nicobar Islands, where there was once a European settlement, some very superior shells have been collected; but since the time the English left the place, no more have been received in this country: they are consequently in very great request.

THE same may also be said of the Andamans, from whence very fine and rare cones, limpets, and chitons (boat-like shells), which adhere to the rocks, have been brought.

A SMALL thorny shell, *(nerite)*, resembling a whelk, of a black unsightly appearance, is found on the coasts which form the bay of Bengal and the entrances of the Ganges; this shell is in request: but there must be many interesting varieties on these shores, as well as land and fresh-water shells, from the interior, of which we at present know very little.

A FEW years ago I received, by the kind remembrance of a gentleman, some very fine small snail-shells, *(helices)*, from Seringapatam, which are the only varieties of this species hitherto known.

MADRAS presents such a surf-beaten coast, that no perfect shells are found there; but many fine varieties, which were sent from Tranquebar, a Danish settlement, have enriched the cabinets of Europe.

WE now come to the famed island of Ceylon[*], well known to Conchologists, for the *rare vo-*

---

[*] When Ceylon was taken, in the year 1795, the collection belonging to the Dutch Governor was purchased for me. It contained some of the finest shells that ever came to England. During the last ten years, I do not hesitate to say, that L.10,000 worth of shells have been sent from this island, more than two-thirds of which have passed through my hands. The *natives* who make up collections in fine partitioned boxes, scarcely ever put a good shell into them: the best shells they sell alone. A peculiar shell from these seas is held in great estimation in China, and is sometimes mounted in pure gold: it is reputed to add great virtue to medicine administered in it!

lutes found on its coast, and for the land and fresh water shells from the interior. The divers employed in the pearl-fishery bring up fine and numerous varieties.

On the Malabar coast, at Tellicherry, I picked up some beautiful sea-weed, and a few small cowries of little importance. At Old-woman Island, near Bombay, I found a fine, though small, Weaver's-shuttle, (*bulla volva*), which shows that it is an inhabitant of the Indian seas.

Hence, until we approach the Persian Gulph, I am not aware that any shells, worth notice, are to be found. From the sands and shores of the adjacent coast, many extremely fine varieties have been gathered, which bear the distinguished names of the *Persian Crown, Voluta Gambronica*, &c.

The coasts of the great island Madagascar abound in shells, but they are generally large, and of little value: some rare varieties are, however, occasionally found. Of the land-shells we are quite ignorant; we are therefore anxious to obtain them. A ship, in which the Author was an officer, sounded on a coral reef of great extent, in the channel of Mosambique, from whence, and nearer the island, he procured some fine specimens.

The Red Sea and its islands produce many fine shells. Lord Valentia, (now Earl Mountnorris), during his travels in those parts, discovered some new varieties. I take this opportunity of acknowledging his Lordship's generosity, in presenting me with his duplicates.

We know nothing of the shells that occur on the eastern coast of Africa, until we arrive at Zanzibar and Mosambique: from these coasts we have received a few interesting varieties, and immense quantities of the commonest class of cowries, which are brought home by the ships that go on these coasts for *right* whales. A few rare fresh water and inland productions, as well as corals, have also been collected. The jaws of the whale in these seas are often covered with curious barnacles; numerous chitons may be found, with limpets, amongst the rocks.

THE Comora islands, particularly Johanna, abound in common cowries. I have seen large heaps of them shovelled up at low water; and, as a peculiar variety, which is found here, passes for currency in Africa, small vessels take in or load considerable quantites of them.

THE Isles of Bourbon and France are highly and deservedly celebrated for shells—and it may be remarked, that whatever is produced there, is the most beautiful of its species. A curious distorted land-shell, which is scarce and extremely interesting, is peculiar to these islands. The officers sometimes amuse themselves in fishing for these beautiful productions; both ladies and gentlemen from thence have made considerable profit by the shells they brought with them.

THE ship which took out the first settlers to Algoa-bay, on her return home, brought me many interesting land and fresh water shells, which the commander was kind enough to collect. On the coast about the Cape, as well as on the rocks and islands in the bays, some good varieties have been found, particularly limpets and large zebra-striped land-shells, which are tender, and require great care in packing.

I HAVE now taken the navigator through most of the seas, and have briefly enumerated the places where he is most likely to discover fine and rare shells. I shall conclude this part of my subject in his own language, advising him to keep a *good look out*.

E

AMBERGRIS.

As this substance is a marine production, I have thought proper to introduce some account of it here.

This delicious perfume is, without doubt, the produce of the sperm whale*, and probably is the result of a disease in the digestive organs. Captain Poole, in pushing a lance through the blubber, and near the passage of the abdomen, felt it strike against something hard, which he thought was a stone, and in drawing it out, the edge was found to be broken. Having cut into the intestine canal, he perceived that the lance

---

* The fibres of the skin from the head of this whale, are of great tenacity; excellent ropes may therefore be manufactured of them, especially for situations exposed to much friction: it is said that they are stronger than catgut, and make better fiddle-strings.

had struck against two large pieces of Ambergris. There is every reason to suppose, that this substance sometimes stops up the passage of the abdomen, and ultimately occasions the death of the animal.

Many other instances have have been recorded of Ambergris having been found in the sperm-whale, and also on the coast, after a sick fish has been seen near the shore.

J.MAWE,149,Strand.

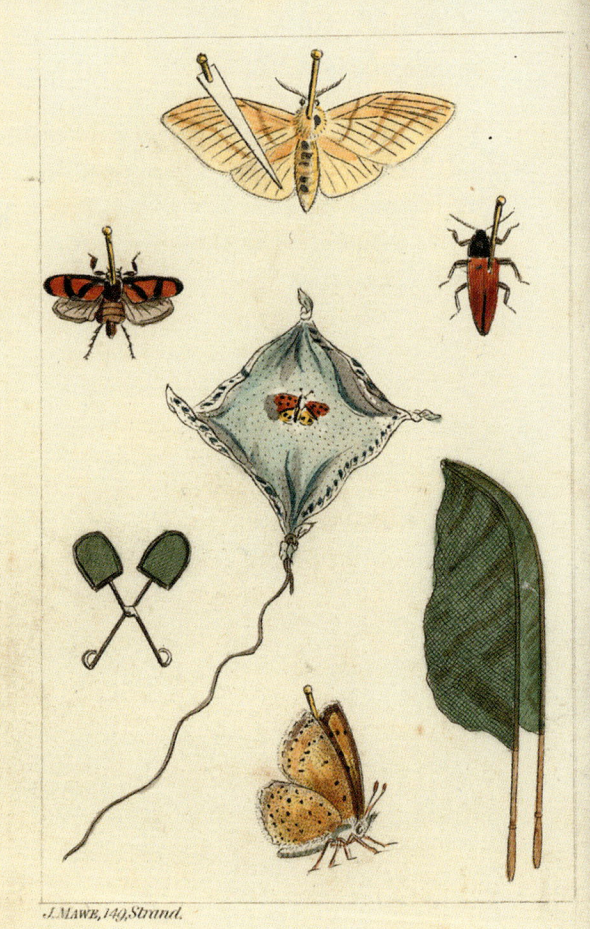

CHAPTER III.

On Insects.

THIS branch of natural history has received considerable attention; but to obtain even a slight knowledge of the subject, much time and great labor are required. The collecting and preserving of insects, is also attended with no small trouble; and what is still worse, it seldom repays those who collect with a view to profit.

We shall proceed to describe the implements that are used, and the methods that are adopted in catching insects at rest or on the wing; but first, let me advise the collector to handle them as little as possible, least he should disturb or destroy the delicate down, to which many of them owe their greatest beauty. Before he proceeds on his search, he will do well to provide himself with a stock of pins, with which he is to pierce the insects he may catch, and a small box lined with cork, or soft wood. With a pair of gauze forceps he may catch insects when at rest; but if they are on the wing, and within reach, he must use a hand-net, which may be made of any light substance, as a piece of gauze about a yard and a half square, fastened to two pliable sticks or canes, with which it may be made to open or collapse at pleasure. If they are beyond his reach, he must use a casting net, which I have tried with considerable success. It may be made thus: tie a weight, (a halfpenny for instance), in one of the corners of a piece of gauze, (about the size of a common handkerchief), a sixpence in the second corner, and a bit of very light wood in the third: the

inequality in the weight and bulk of these substances, will occasion the gauze to open when thrown from the hand: a thin piece of twine, a yard or two long, may be tied to the remaining corner, by which the net may be drawn in at pleasure. The art of spreading it to its full extent may be acquired with very little practice.

Having caught the insects, the next thing is to preserve them. Moths, butterflies, locusts, and others of this class, may be killed by nipping them across the thorax. Wasps, bees, hornets, &c. when secured, may be treated in the same way, guarding the hand with a handkerchief; or they may be squeezed with a pair of forceps: but if the collector be not careful in performing this operation, he will in all probability have cause to regret his want of caution. Or they may be killed by putting them in a glass immersed half way up in boiling water, and covering the top close; or by placing them on a plate under an inverted tumbler, and setting it before the fire for a minute or two. I have known gentlemen to put colleopterous insects, as beetles, wasps, &c. into a common pocket bottle half full of spirits, with which they have travelled some days, and brought them home quite perfect.

THE intestines of butterflies and large insects should be extracted, which may be done by cutting a slit with a fine-pointed pair of scissars, at the extremity of the body, and gently pressing them out; a small roll of cotton or paper, dipped in the preventive soap, should then be introduced, so as to extend the body to its natural form.

INSECTS have been frequently rendered less interesting, by packing them in cotton, which is perhaps one of the worst substances that can be used for this purpose, as the very delicate claws, feelers &c. of some species, are certain to become entangled in it. The best method is to stick the pins (on which they are fixed) very fast into the bottom, sides, and top of the box I have before described. When the box is full, and the insects quite dry, a small portion of camphor should be placed securely in the corners, and the openings should be closed with pitched canvass, otherwise the ants, which are so numerous in hot countries, would enter and devour the contents.

THE finest insects are brought from the tropical climates. Brazil, India, Java, China, &c. produce beautiful species.

THE insects from new countries, and those islands and remote parts which are seldom visited, cannot fail of exciting interest, either by their beauty or rarity.

CHAPTER IV.

On Birds.

MANY species of Birds, of surpassing beauty, have been brought by travellers from foreign countries, and have been domesticated here; and the skins of others, (whose tender nature unfits them for our colder climate), have been brought home and preserved. Thus, to a great extent, we possess the fine varieties of the feathered creation, belonging to tropical climates.

THE process to be observed in taking the skin from birds, is not at all difficult; but it would greatly facilitate the acquiring dexterity in the art, to see the operation once or twice performed by a skilful practitioner.

Before the operator proceeds to remove the skin, he should place the plumage as smooth as possible, and carefully clean it from any spot of blood or dirt that may appear upon it: a little soft linen rag, or paper, should be placed in the mouth, which should then be sewed or tied up, to prevent any blood issuing from it. The bird should now be laid on its back, and an incision made with a knife along the breast bone, (where the feathers divide), as far as the vent, taking great care not to cut the flesh: an ivory paper-knife, or the fingers, may then be introduced, to separate the skin from the breast *. The thighs being gently forced up, the flesh should be cut off, leaving the bone quite clean. The skin may now be easily separated from the body down to the rump, which must be cut off; then draw it over the back, as far as the wings, which cut

---

* Some absorbent, as chalk or flour, should be applied occasionally to the inside of the skin, to prevent its adhering to the body.

off close to the body; then, pushing the joint from the outside inwards, the skin will easily separate from the flesh, which must be scraped off the bones. It may then be pulled over the neck and part of the head, as far as the eyes; the head must be pressed inwards and separated from the neck. In the back of the skull a hole should be made with a penknife, of sufficient size to admit an instrument resembling an ear picker, but rather larger, with which the brains, eyes, tongue, and the fleshy membranes may be extracted, taking care not to disturb the bones of the head. The skin must be very carefully examined, (particularly about the vent and rump) and every piece of flesh or fat removed, and the feathers placed in exact order.

As the skin in this state is extremely susceptible of enlargement, the greatest precaution must be used not to extend it. The preservative powder, or soap, is now to be carefully introduced into every part; and if any appearance of moisture remains, it should be completely absorbed by a linen cloth, and the powder again applied. It is necessary to observe, that the skin must not be hung up to dry, unless a string is passed under it, from the rump to the head, so that it may hang on the string, and not by any part of the skin, which would otherwise stretch beyond its natural size.

The skin may now be prepared for packing, by placing the wings and extremities in their proper positions, and laying smooth the plumage: a little cotton may be put in the inside, and sewed up, to preserve the form, after which it should be carefully folded in paper, or placed between the leaves of a book, and kept free from damp.

CHAPTER V.

On Reptiles.

IN treating upon this subject, I shall principally confine myself to the methods that may be employed in catching and skinning reptiles.

The fangs with which they defend themselves, or attack their victims, are, in venomous serpents, placed in the outside of the jaw, and so fixed, that they may be erected or depressed at pleasure; they are mostly from half an inch to three quarters long, with a very small slit at the point, and generally a little crooked.

SERPENTS may be caught with a wire-noose, fixed to the end of a pole, by passing it over their heads. Thus I have brought them to a convenient place; and, with a pair of forceps, a bit of pointed wood may be introduced into their mouths, to extend their jaws, in order to examine them. They are easily killed by a slight blow on the head.

THE best method of taking off the skin, is to make an incision at the vent, a few inches in length, or even up to the head. The skin may then be separated, by introducing the fingers betwixt it and the body; or, if the scales are not large, it may be skinned in the same manner as is commonly practised with eels. The body must then be cut off from the head, and the brains, together with the eyes, and all the fleshy parts must be taken out, without disturbing the fangs, jaws, or tongue. The skin may then be pulled down as far as the tail, which should be cut off an inch or two from the extremity. The body may be preserved in spirits, in order to show the moveable ribs and flexibility of the spine.

THE skin, thus freed from the body, must be examined and cleared from flesh and fat, and the head cleaned as well as possible. The preservative powder may be used where any muscular ligaments or flesh remain, and the soap may be applied to the skin, which must then be hung up in the air. If, after a day or two, any moisture should appear, it must be absorbed by a cloth, and more powder applied, until every part is dry. In these operations, the scales, &c. should be attended to, and if displaced by skinning or otherwise, they should be pressed into their exact position, before the skin is perfectly dry.

LIZARDS, alligators, frogs, &c. &c. may be

treated in the same manner. When the operation is completed, the skin should be rolled up, and packed securely in paper, and afterwards sewed up in canvass.

THE spine of a common sized serpent may be easily broken with a sharp blow from a stick; after which, I have witnessed them very vicious, boldly attacking and biting whatever was opposed to them.

I ONCE drew a rattle-snake to a hog, which devoured it, notwithstanding its bite.

VERY large serpents, such as are thirty or forty feet long, are formidable beyond any ideas we can entertain. I have found it difficult to disengage one, not even four feet in length, that had coiled itself round my arm, when suspended by a wire to a nail in my door.

CHAPTER VI.

On Quadrupeds.

AS it is often impossible to convey home the quadrupeds a traveller may meet with in visiting foreign lands, it becomes necessary to take off and preserve their skins; many of which, especially those from newly discovered countries, interest us extremely, either by their beauty or their novelty.

THERE are few who have not seen a poulterer take the skin from a rabbit, or a butcher perform the same operation on a sheep or ox: a lesson from either of these persons would be of use to the traveller. In taking the skin from large or small animals, the same process is required. The operation must commence by making a slit, of any length, along the belly, so as to give the greatest facility in using the fingers, or a knife, to separate the skin from the muscles, &c. The legs of small animals may be pressed inwards, and the skin stripped over them, inside out, great care being taken in passing the joints: the feet and tail must be preserved as well as possible, by cutting out the flesh, and preserving the hoofs, nails, or claws. The skin, freed from the extremities, and separated from the abdomen, may be stripped over the back to the neck. Particular attention will be required in stripping it from the neck to the ears and the nose, from the latter of which it must be cut off at the termination of the bone, carefully preserving the extremity. The skin, if necessary, may be cut under the jaw, as it can afterwards be sewed up. The head of the animal may now be cut off, at the back of which a hole must be made to extract the brains; it being desirable, in small animals, to disturb the skull bones as little as possible. The ears may be cut off close to the head, and afterwards cleaned. The eyelids, lips, jaws, and teeth must be preserved; and every thing done to make the animal appear as if alive.

It is well to keep the carcass as whole as possible during the operation, to prevent the flowing of blood, &c. which would prove troublesome to the operator. Towels and saw-dust should be at hand, to be used when wanted.

The skin, now free, must be wiped clean, the membraneous and fleshy parts taken away, and the extremities, (particularly the head), kept as entire as possible, especially in small animals. The skin being now perfectly cleaned, may have the preservative soap and powder applied all over it: flax or bits of rag, well anointed with the soap, may be placed in the head, nostrils, and about the hoofs, claws, and tail. It should then be laid out to dry for a day or two; and, on a second examination, where any moisture appears, it must be absorbed by a cloth, and more powder applied, until it is quite dry.

The skin may now be stuffed with cotton, &c. and sewn up, to keep it in some degree in its natural form: or it may be rolled up and packed in canvass, and stowed away in a case or barrel. The hides of large animals, as oxen, seals, &c. &c. are often brought from remote parts with no other preparation than salt.

When the preservative powder is used, the operator should be very careful not to leave any of it about, least a domestic animal should get to it and eat it. I once lost a beautiful monkey, which was poisoned by licking up a portion that was incautiously left.

The learner will do well to practise on a squirrel, rabbit, fox, &c. and then he will be prepared for the practical difficulties he may meet with in taking the skin from the feet, head, and tail of other animals. The remaining parts of the operation are very easily performed.

CHAPTER VII.

On Plants, &c.

THERE is scarcely a more interesting science than that of Botany; and though the traveller may not be a professed botanist, yet he may be pleased with the endless variety of beautiful plants and flowers which every country produces, or gratified with the opportunity of enriching our gardens and fields by the introduction of new varieties. How much has our agricultural interest been benefitted by the inportation of varieties of grass, as lucern, clover, &c. and what do we not owe to him who first brought hither the potatoe? The Japan rose, and many

H

other exotics agree so well with this climate, that they may almost be said to be naturalized. The voyager, therefore, in distant climes, should not disregard any vegetable production. The corn, pulse, and roots that are used abroad in domestic economy, are highly worthy his attention, in a commercial point of view. There are two public institutions in this country for the reception of whatever is brought from abroad. The board of agriculture, for grass and whatever belongs to the economy of animals or man; and the horticultural society, for seeds or cuttings of fruit trees, exotic plants, &c. where every attention will be paid to their growth and culture.

Woods, bark, (dye woods), many are highly valuable in commerce, and much in request.

Lichens, (moss), some produce fine and permanent dyes, as the orchella, and are very valuable; others are medicinal, and continually in request.

Gums—Their general use and value are well known; they always form an article of commerce, and are used for an infinite number of purposes.

Seeds of every sort, and any remarks relative to the plant, will be interesting. They must be gathered and kept dry.

Plants, leaves, or flowers, may be preserved between the leaves of a book, forming a *hortus-siccus*: these are collected and preserved with very little trouble, and may be considered in two points of view:—First, as an agreeable amusement; and next, as giving that information which may become highly beneficial to society.

CHAPTER VIII.

On Minerals.

IT may be necessary to say something on minerals, in the pursuit of which the greater part of my life has been employed. A traveller who is unacquainted with metals, should procure a few in a rough state, and, by comparison, he would soon know how to discriminate one from the other. A small book which the author published, called "FAMILIAR LESSONS ON MINERALOGY," with colored plates, would greatly facilitate his inquiries, and cannot be too strongly recommended to those who are desirous to know any thing of minerals.

Pieces of rocks, with the names of the places from whence they came, would be always interesting, as we are unacquainted whether many islands, head-lands, &c. are granite, limestone, or volcanic. Collections of rocks, with particulars concerning them, are highly desirable, in order to determine the relative connection of mountains, islands, &c.

The soil at the bottom of streams or rivers, if gravel, generally partakes of interesting subjects. In India, precious stones occur in such soil: in Africa and South America, gold, platina, diamonds, rubies, sapphires, and topazes, belong to, and are always found in gravel beds, alluvial soil, as is Tin, in the island of Banca. Silver, lead, copper, iron, &c. occur in veins.

Wherever there are mines, (subterraneous excavations), metals of some sort or other are produced; and it surely is not burthensome to the intellect to know lead ore from copper, or silver from lead, or gold from iron, or diamonds from pebbles.

Permit me to advise the traveller to look into the book of nature, which is always open, and learn what he can. A little information on this head may prove highly advantageous, as the wealth of nations mostly depends on the produce of their mines. It is earnestly to be recommended, wherever he goes, to bring from thence some of the rocky substances, and if any other present themselves, he should endeavour to possess himself of them, which he might examine at leisure, with the simple instrument the blowpipe, the use and mode of managing which is explained in a small Treatise* intended to accompany the Lessons on Mineralogy.

* How to detect gold when adulterated, is particularly explained; the book will be found useful to those who go on the coast of Africa, South America, China, &c.

FINIS.

W. M'Dowall, Printer, Pemberton Row,
Gough Square.

REFERENCES AND FURTHER READING

The Shell Collector's Pilot

Mawe, John. 1821. *The Voyager's Companion; or Shell Collector's Pilot: with Instructions and Directions where to find the finest shells; also for preserving the skins of animals; and the best methods of catching and preserving insects, &c. &c. &c.* 3rd edition, printed and sold by the Author, 149, Strand, and Longman, Hurst, Rees, Orme & Brown, Paternoster Row, London, coloured plates, xv + 56 pp., royal 16mo, bound in half-calf and marbled boards with dark red morocco spine label and title in gilt, or paper covered boards.

1825, 4th and last edition, published by the Author, London, vii + 75 pp., with one-page advertisement, usually bound in dark green pebbled cloth or salmon-coloured paper covered boards with title on paper label on front cover.

Only one copy is recorded of the first two editions: 1804, *A Short Treatise, addressed to Gentlemen visiting the South Seas, and all Foreign Countries; more particularly to Commanders, &c. of Ships, and Gentlemen residing on Shore, with a view to encourage the collecting of Natural History…*, 1st edition, published by Young, Brydges Street, Covent Garden, London, unillustrated in wrappers, 11 pp.

No copies of the second edition have been located, and the date of publication is unknown.

Library sources

British Library, Manuscript Collections, 96 Euston Road, London

Geological Society of London, Burlington House, Piccadilly, London

Natural History Museum, London, Official Archives (General Library, Earth Sciences Library)

Cited works and further reading

Abbott, R. Tucker. 1972. *Kingdom of the Seashell.* Crown Publishers, New York, 256 pp.

Dance, S. Peter. 1986. *A History of Shell Collecting.* E.J. Brill / Dr W. Backhuys, Leiden, 265 pp. (originally published by Faber & Faber, 1966).

Edwards, W.N. 1967. *The Early History of Palaeontology.* The British Museum (Natural History), London, Publication No. 658, 59 pp.

Hoare, Sarah. 1831. *Poems on Conchology and Botany, with plates and notes.* Simkin & Marshall, London, 106 pp.

Lister, Martin. 1678. *Historiae Animalium Angliae tres Tractatus unus de Araneis alter de Cochleis tum terrestibus tum fluviatilibus tertius de Cochleis Marinis. Quibus adjectus est Quartus de Lapidibus ejusdem insulae ad cochlearum quandam imaginem figuratis Memoriae & Rationi.* Joh. Martyn, London, 250 pp. plus Errata page, plates.

Mawe, John. 1802. *The Mineralogy of Derbyshire: with a description of the most interesting mines in the North of England, in Scotland, and in Wales…* William Phillips, Derby, 211 pp., plates.

Mawe, John. 1813. *A Treatise on Diamonds and Precious Stones, including their History—Natural and Commercial, to which is added some account of the best methods of cutting and polishing them.* Longman et al., London, xii + 166 pp., plates (2nd edition 1823).

Mawe, John. 1815. *Travels in the Interior of Brazil, particularly in the Gold and Diamond District of that fine country…* Longman et al., London, vii + 366 pp., plates (other editions include earlier edition and 1816 edition published in Philadelphia, 1823 and 1825 editions in London, some undated editions).

Mawe, John. 1816. *A Descriptive Catalogue of Minerals, intended for the Use of Students…* 2nd edition, published by the Author, London, xiv + 94 pp., frontispiece (very rare 1st edition 1815).

Mawe, John. 1820. *First Lessons on Mineralogy and Geology…* 2nd edition, Longman et al., London, viii + 96 pp., plates (rare 1st edition 1819, last edition 1832 published by Sarah Mawe).

Mawe, John. 1823. *The Linnaean System of Conchology, describing the Orders, Genera, and Species of Shells, arranged into Divisions and Families.* Longman et al., London, xv + 207 pp., plates by Crouch.

Mawe, John. 1832. *Wodarch's Introduction to the Study of Conchology: describing the Orders, Genera, and Species of shells with observations on the nature and properties of the animals; and directions for collecting, preserving, and cleaning shells.* Longman et al., London, and Sarah Mawe, 149 pp., 6 plates (1st edition by Wodarch 1820, 2nd edition by Mawe 1822, London, 152 pp.).

Melvin, A. Gordon. 1973. *Seashell Parade: Fascinating Facts, Pictures and Stories.* Charles E. Tuttle Co., Rutland, Vermont, USA and Tokyo, Japan, 369 pp.

Pinnock's Catechisms. 1824. *A Catechism of Conchology, containing a pleasing and familiar description of the Construction and Classification of Shells, according to the Linnean Systeml with examples of each class.* G. and W.B. Whittaker, London, 71 pp. plus index (2nd edition 1829).

Roberts, Mary. 1834. *The Conchologist's Companion.* Whittaker & Co., London, 210 pp.

Rosenberg, G. 1992. *The Encyclopedia of Seashells.* Dorset Press, New York, 224 pp.

Rumphius, Georgius Everhardus. 1739. *Thesaurus Imaginum Piscium Testaceorum...* Hagae-Comitum, Petrum de Hondt, 14 pp., 60 plates (see also 1705 and 1766 works by Rumphius in Dance, 1986).

Saul, Mary. 1974. *Shells: An Illustrated Guide to a Timeless and Fascinating World.* Country Life, London, 192 pp.

Say, Thomas. 1819. *Land and Fresh Water Shells of the United States.* 3rd edition, Philadelphia, 20 pp. including 4 plates (1st edition 1816 published in Mitchell, Ames and White, *Nicholson's Encyclopedia*; printed separately 1817 as *Descriptions of Land and Fresh-Water Shells of the United States*; 2nd edition 1818).

Stix, Hugh and Marguerite, and Abbott, R. Tucker. 1972. *The Shell: 500 m.y. of Inspired Design.* Ballantine Books, New York, unpaginated.

Swainson, William. 1840. *Taxidermy; with the biography of zoologists.* Longman et al., London, 393 pp. Larder's Cabinet Cyclopaedia, pp. 270–1.

Tomlin, J.R. Le B. 1943. Book notes. 10. John Mawe. *Linnaean System of Conchology. Proceedings of the Malacological Society of London,* 25: 141.

Torrens, Hugh S. 1992. The early life and geological work of John Mawe 1766–1829 and a note on his travels in Brazil. *Bulletin of the Peak District Mines Historical Society,* 11 (6): 267–71, 282.

Wye, Kenneth R. 1991. *The Illustrated Encyclopaedia of Shells...* Headline, London, 288 pp.

Zahl, Paul A. 1969. The magic lure of sea shells. *National Geographic,* 135 (3): 386–429.